Semigroups and Their Applications

Semigroups and Their Applications

Proceedings of the International Conference "Algebraic Theory of Semigroups and Its Applications" held at the California State University, Chico, April 10-12, 1986

edited by

Simon M. Goberstein

Department of Mathematics, California State University, Chico, U.S.A.

and

Peter M. Higgins

Department of Mathematics, California State University, Chico and Division of Computing and Mathematics, Deakin University, U.S.A.

D. REIDEL PUBLISHING COMPANY

A MEMBER OF THE KLUWER ACADEMIC PUBLISHERS GROUP

DORDRECHT / BOSTON / LANCASTER / TOKYO

Library of Congress Cataloging in Publication Data

CIP

International Conference, "Algebraic Theory of Semigroups and Its Applications" (1986: California State University, Chico)
Semigroups and their applications.

Includes index.
1. Semigroups—Congresses. I. Goberstein, Simon M., 1949–
II. Higgins, Peter M., 1956– .III. Title.
QA171.I59 1986 512'.2 87-4591
ISBN 90-277-2463-6

Published by D. Reidel Publishing Company,
P.O. Box 17, 3300 AA Dordrecht, Holland.

Sold and distributed in the U.S.A. and Canada
by Kluwer Academic Publishers,
101 Philip Drive, Assinippi Park, Norwell, MA 02061, U.S.A.

In all other countries, sold and distributed
by Kluwer Academic Publishers Group,
P.O. Box 322, 3300 AH Dordrecht, Holland.

Printed in The Netherlands

TABLE OF CONTENTS

PREFACE

Most papers published in this volume are based on lectures presented at the Chico Conference on Semigroups held on the Chico campus of the California State University on April 10-12, 1986. The conference was sponsored by the California State University, Chico in cooperation with the Engineering Computer Sciences Department of the Pacific Gas and Electric Company. The program included seven 50-minute addresses and seventeen 30-minute lectures. Speakers were invited by the organizing committee consisting of S. M. Goberstein and P. M. Higgins.

The purpose of the conference was to bring together some of the leading researchers in the area of semigroup theory for a discussion of major recent developments in the field. The algebraic theory of semigroups is growing so rapidly and new important results are being produced at such a rate that the need for another meeting was well-justified. It was hoped that the conference would help to disseminate new results more rapidly among those working in semigroups and related areas and that the exchange of ideas would stimulate research in the subject even further. These hopes were realized beyond all expectations. All talks presented at the conference were of high quality; together, they covered a vast array of topics: semigroup algebras, transformation semigroups, varieties and pseudovarieties of some classes of semigroups, global semigroup theory, various topics in the theory of inverse semigroups, semigroups presented by relations, and connections with languages and automata. An enlightening lecture presented at the conference by T. E. Hall about C. Ash's brilliant result that every finite monoid with commuting idempotents is a morphic image of a subsemigroup of a finite inverse semigroup stimulated J.-C. Birget, S. Margolis and J. Rhodes to extend Ash's ideas and to prove that every finite semigroup whose idempotents form a subsemigroup divides a finite orthodox semigroup. These results show that the so-called "Type II Conjecture" stated by J. Rhodes and B. Tilson in 1972 holds for finite monoids whose idempotents form a submonoid. B. Tilson wrote an article, specifically for this volume, clarifying the proof of the original result that led to the Type II Conjecture. This article is included in the Proceedings even though it was not presented at the conference.

The papers in this volume are written in a variety of forms. Some represent extensive surveys, others are written as regular research articles. A number of contributions are research announcements whose proofs should appear elsewhere. And several papers represent a mixture of all three categories of articles. There is no need to discuss here the contents of each contribution. The reader is advised to read them all.

This conference would never have taken place if it had not been generously financed by the Office of the Provost, the Graduate School and the College of Natural Sciences of the California State University, Chico and the Pacific Gas and Electric Company. The support of Provost G. R. Stairs and Associate Vice President for Research J. S. Morgan was especially vital. R. J. Bakke from the Graduate School worked hard trying to raise external funds to support the conference and B. L. Lundy from the College of Natural Sciences did a lot of work as the conference accountant. T. A. McCready, Chairman of the Department of Mathematics, constantly encouraged and supported me during the preparation of the conference and the editing of its Proceedings. A. Bowman, S. Jones and B. Stansbury, Mathematics Department secretaries, worked tirelessly typing various materials related to the conference as well as parts of this volume. The work and help of all these individuals is greatly appreciated.

Shortly after the first announcement about the conference had been mailed, my colleague and friend, P. M. Higgins, left Chico to take an appointment at Deakin University. In spite of the great distance between Chico and Geelong, Peter's interest in the conference never weakened. He arrived in Chico a week before the meeting and was eager to help with the final preparations. For all that and for his willingness to help me with editing some of the papers for this volume I owe him my gratitude.

Last but not least I am especially grateful to my wife Faina for her encouragement and understanding during the preparation of the meeting and the editing of this volume and for being such a gracious hostess to the conference participants.

Chico, California, December 26, 1986 Simon M. Goberstein

LIST OF PARTICIPANTS

J. Almeida (Braga, Portugal)

J.-C. Birget (Lincoln, Nebraska)

R. Brandon (La Grande, Oregon)

S. Bulman-Fleming (Waterloo, Ontario)

K. Byleen (Milwaukee, Wisconsin)

D. Cowan (Burnaby, British Columbia)

M. Drazin (West Lafayette, Indiana)

M.D. Gass (Collegeville, Minnesota)

J.A. Gerhard (Winnipeg, Manitoba)

S.M. Goberstein (Chico, California)

M. Gould (Nashville, Tennessee)

T.E. Hall (Clayton, Victoria)

H. Hamilton (Sacramento, California)

D. Hardy (Fort Collins, Colorado)

K. Henckell (Willits, California)

P.M. Higgins (Chico, California and Geelong, Victoria)

J.A. Hildebrant (Baton Rouge, Louisiana)

J.M. Howie (St. Andrews, Scotland)

J. Iskra (Macon, Georgia)

P.R. Jones (Milwaukee, Wisconsin)

K. Johnston (Charleston, South Carolina)

G. Karpilovsky (Johannesburg, South Africa)

R.J. Koch (Baton Rouge, Louisiana)

W. Lex (Clausthal, West Germany)

K.D. Magill (Buffalo, New York)

E. Manes (Amherst, Massachusetts)

S. Margolis (Lincoln, Nebraska)

K. McDowell (Waterloo, Ontario)

J. Meakin (Lincoln, Nebraska)

D.W. Miller (Lincoln, Nebraska)

P. Misra (Staten Island, New York)

W.D. Munn (Glasgow, Scotland)

G. Naber (Chico, California)

W. Nico (Hayward, California)

T.D. Parsons (Chico, California)

F. Pastijn (Milwaukee, Wisconsin)

M. Petrich (Burnaby, British Columbia)

J.-E. Pin (Paris, France)

N.R. Reilly (Burnaby, British Columbia)

L. Renner (London, Ontario)

J. Rhodes (Berkeley, California)

H.E. Scheiblich (Columbia, South Carolina)

R. Spake (Davis, California)

J. Stephen (Lincoln, Nebraska)

H. Straubing (Boston, Massachusetts)

T. Tamura (Davis, California)

M.C. Thornton (Lincoln, Nebraska)

B. Tilson (Sausalito, California)

E. Vought (Chico, California)

S. Wismath (Lethbridge, Alberta)

K. Yamaoka (Davis, California)

SCIENTIFIC PROGRAM

Thursday, April 10

Morning session (Chairman - S.M. Goberstein):

Norman Reilly, Simon Fraser University (Canada)
'The lattice of varieties of completely regular semigroups'
(a 50-minute address)

Jorge Almeida, Universidade do Minho (Portugal)
'Implicit operations on certain classes of semigroups'

William Nico, California State University, Hayward
'Categorical extension theory - revisited'

Francis Pastijn, Marquette University
'Uniform chains'

Afternoon session (Chairman - J. Rhodes):

W.D. Munn, University of Glasgow (Scotland)
'A class of inverse semigroup rings' (a 50-minute address)

K. Johnston, College of Charleston
'Modular inverse semigroups'

T. Tamura, University of California, Davis
'On the recent study of power semigroups'

J.-E. Pin, Université Pierre et Marie Curie-Paris VI (France)
'Finite power semigroups: A survey'

J.A. Gerhard, University of Manitoba (Canada)
'Some free semigroups and free *-semigroups'

Friday, April 11

Morning session (Chairman - P.M. Higgins):

John Rhodes, University of California, Berkeley
'New techniques in global semigroup theory' (a 50-minute address)

Stuart Margolis, University of Nebraska-Lincoln
'Word problems for inverse semigroups'

J.C. Meakin, University of Nebraska-Lincoln
'The E-unitary problem for one-relator inverse monoids'

W. Lex, Technische Universität Clausthal (F.R.G.)
'Lattices of torsion theories for semi-automata'

Afternoon session (Chairman - R.J. Koch):

T.E. Hall, Monash University (Australia)
'Ash's theorem: any finite semigroup with commuting idempotents divides a finite inverse semigroup' (a 50-minute address)

Karl Byleen, Marquette University
'Inverse semigroups with countable universal semilattices'

J.-C. Birget, University of Nebraska-Lincoln
'Global theory of semigroups whose idempotents commute'

Mario Petrich, Simon Fraser University (Canada)
'Cayley theorems for semigroups' (a 50-minute colloquium address)

Saturday, April 12

Morning session (Chairman - J.M. Howie):

Peter Jones, Marquette University
'Basis properties, exchange properties and embeddings in idempotent-free semigroups' (a 50-minute address)

M.P. Drazin, Purdue University
'Units of inverse semigroup algebras'

K.D. Magill, SUNY at Buffalo
'The countability indices of certain transformation semigroups'

Matthew Gould, Vanderbilt University
'Cofinality of normal bands of groups'

Afternoon session (Chairman - W.D. Munn):

T.E. Hall, Monash University (Australia)
'The amalgamation bases of the class of finite inverse semigroups' (a 50-minute address)

Howard Straubing, Boston College
'Partially ordered finite monoids and a theorem of I. Simon'

J.M. Howie, University of St. Andrews (Scotland)
'Rank properties in semigroups of mappings'

IMPLICIT OPERATIONS ON CERTAIN CLASSES OF SEMIGROUPS

Jorge Almeida* and Assis Azevedo*
Centro de Matemática
Universidade do Minho
Braga, Portugal

ABSTRACT. Varieties of algebras are characterized by identities, where an identity is a formal equality of two terms (i.e., operations defined by means of the underlying operations). Analogously, pseudovarieties of (finite) algebras are defined by pseudo-identities, these being formal equalities of so-called implicit operations (briefly , functions compatible with all homomorphisms). To further explore this analogy to yield results on finite algebras, it is necessary to obtain clear descriptions of implicit operations. This work is a contribution to this project in the area of semigroup theory. All unary implicit operations on semigroups are described, and the implicit operations on certain pseudovarieties of semigroups are given in terms of "generating" operations. The existence of some unusual implicit operations is established based on classical combinatorial theorems about words.

1. INTRODUCTION

This paper is concerned with a new way of looking at pseudovarieties of semigroups : via implicit operations. Roughly put, an implicit operation is a new operation which is preserved by all functions that preserve the old operations (i.e., homomorphisms). Reiterman [8] showed that implicit operations on finite algebras of a finite finitary type form a compact metric space in which the subset of finite composites of old operations is dense. He also showed that pseudovarieties are defined by pseudoidentities, i.e., by formal equalities of implicit operations, thus providing a suitable analog to the classical Birkhoff theorem on varieties.

We start here a systematic study of implicit operations on finite semigroups. Our first positive result is a full constructive description of unary implicit operations. This already allows us to show there is a vast unexplored world of implicit operations compared with what can be found in the literature on pseudovarieties (cf. Eilenberg [4] and Pin [7]).

* This work was supported, in part, by INIC grant 85/CEX/04

S. M. Goberstein and P. M. Higgins (eds.), Semigroups and Their Applications, 1–11.

We then proceed to show there is apparently more and more complication as the arity n of implicit operations increases. For $n=2,3$, this is based on classical results on avoidable regularities in words on two and three-letter alphabets.

Finally, we restrict our attention to the class ZE of all finite semigroups in which idempotents are central. We show that implicit operations on finite groups and explicit operations suffice to obtain all implicit operations on ZE . This is based on a careful study of sequences of words depending on some simple algebraic and combinatorial lemmas.

For basic notation on semigroups and pseudovarieties, the reader is referred to Lallement [5].

2. BACKGROUND AND NOTATION

A class V of finite algebras of a given type is said to be a pseudovariety if it is closed under homomorphic images, subalgebras and finitary direct products. We say V is equational if there is a set Σ of identities such that V is the class of all finite algebras of the given type which satisfy all the identities in Σ .

EXAMPLE 2.1. The pseudovariety N of all finite nilpotent semigroups is not equational since N satisfies no nontrivial semigroup identities, i.e., the least equational pseudovariety containing N is the class S of all finite semigroups. Thus, identities do not suffice to define pseudovarieties.

Let C be any class of algebras of a given type. An n-ary implicit operation π on C associates with each $A \in C$ an n-ary operation $\pi_A : A^n \longrightarrow A$ in such a way that if $A,B \in C$ and $\varphi : A \longrightarrow B$ is a homomorphism then $\pi_B(\varphi(a_1),\dots,\varphi(a_n)) = \varphi(\pi_A(a_1,\dots,a_n))$ for all $a_1,\dots,a_n \in A$. The class of all n-ary implicit operations on C is denoted by $\overline{\Omega}_n C$.

For $A \in C$ and $\pi,\rho \in \overline{\Omega}_n C$, we write $A \models \pi = \rho$ to mean $\pi_A = \rho_A$; we call $\pi = \rho$ a pseudoidentity for C and we then say A satisfies this pseudoidentity. If Σ is a set of pseudoidentities for C, $A \models \Sigma$ means A satisfies all the pseudoidentities in Σ and $[\![\Sigma]\!]$ denotes the class of all $A \in C$ such that $A \models \Sigma$.

If $t(x_1,\dots,x_n)$ is a term of a given type τ in the variables $x_1,\dots,x_n$, then t defines an n-ary implicit operation on any class C of algebras of type τ by letting t_A be the induced n-ary operation on $A \in C$ (where $t_A(a_1,\dots,a_n)$ is obtained by "substituting" $a_i \in A$ for x_i $(i=1,\dots,n)$). An implicit operation of this type is said to be explicit. The set of all n-ary explicit operations on C is represented by $\Omega_n C$.

In $\overline{\Omega}_n C$ it is possible to define a metric distance in the case - which we assume in the following - C is "essentially countable", i.e., up to isomorphism, there are only a countable number of algebras in C. This is the case, for instance, when C consists of finite algebras and the underlying type is finite and finitary. We will not describe here this distance function but rather the convergence of sequences in $\overline{\Omega}_n C$ since this is what will be used in the sequel. A sequence $(\rho_n)_n$ converges in $\overline{\Omega}_m C$ to π if, for every integer k, there is an integer n_k such that $A \models \pi = \rho_n$ for all $n \geq n_k$ and all $A \in C$ with $|A| \leq k$, where $|A|$ denotes the cardinality of A.

Next, we quote two important results of Reiterman [8] .

THEOREM 2.2. $\overline{\Omega}_n C$ is a compact metric space in which $\Omega_n C$ is dense.

THEOREM 2.3. Let $\underline{V}$ be a class of finite algebras of type τ . Then $\underline{V}$ is a pseudovariety if and only if there is a set Σ of pseudoidentities for the class of all finite algebras of type τ such that $\underline{V} = [\![\Sigma]\!]$.

In view of Theorem 2.3, it is natural to study pseudoidentities and implicit operations in order to achieve a better understanding of pseudovarieties.

From here on, we restrict our attention to classes of finite semigroups. For finite semigroups there is one more very natural implicit operation found in the literature besides the explicit operations : the idempotent unary operation x^ω . For an element s of a finite semigroup, the value s^ω of x^ω on s is the idempotent in the subsemigroup generated by s. There is, however, a need for more implicit operations in order to be able to define all pseudovarieties in terms of pseudoidentities.

EXAMPLE 2.4. Let $\underline{Ab}_p$ denote the class of all finite abelian p-groups, and let $\underline{Ab}^p$ denote the class of all finite abelian groups without elements of order p, where p is any prime. We claim $\underline{Ab}_p$ and $\underline{Ab}^p$ cannot be defined by pseudoidentities in which all implicit operations are composites of x^ω and explicit operations.

For, suppose $\underline{Ab}_p = [\![x^\omega = 1, \Sigma]\!]$ where Σ is a set of such pseudoidentities. (In general, $\pi=1$ is an abbreviation of $\pi(x_1,\ldots,x_n)y=y=y\pi(x_1,\ldots,x_n)$.) Then, every $\pi=\rho$ in Σ can be replaced by an identity of the form $v=1$ or $v=w$. Thus, we may assume Σ is a set of identities.

Let $[\Sigma]$ denote the class of all semigroups satisfying the identities in Σ . Of course, $[\Sigma]$ is a variety and the class $[\Sigma]^F$ of all finite members of $[\Sigma]$ is precisely $[\![\Sigma]\!]$. But $\mathbb{Z} \in [\Sigma]$ since $[\![\Sigma]\!]$ contains cyclic groups of arbitrarily large order and $[\Sigma]$ is a variety. Hence $\mathbb{Z}_q \in [\Sigma]^F \cap [\![x^\omega=1]\!] = \underline{Ab}_p$ for all q, contradicting the definition of $\underline{Ab}_p$. A similar argument works for $\underline{Ab}^p$.

3. UNARY IMPLICIT OPERATIONS

In a cyclic semigroup $\langle a \,;\, a^n = a^{n+k} \rangle$, we call n the index of a and k the period of a ; we also denote by K_a its maximal subgroup.

Let $\underline{S}$ denote the class of all finite semigroups.

LEMMA 3.1. Let $\pi \in \overline{\Omega}_1\underline{S}$ be such that $\pi_A(a) \notin K_a$ for some $A \in \underline{S}$ and $a \in A$. Then $\pi \in \Omega_1\underline{S}$.

PROOF. By Theorem 2.3, there is a sequence $(x^{\alpha_n})_n$ of words such that $\lim_{n\to\infty} x^{\alpha_n} = \pi$ in $\overline{\Omega}_1\underline{S}$. Now, if $\alpha_n \geq |A|$, then $\pi_A(a) \in K_a$, so that the set of exponents $\{\alpha_n : n = 1,2,\ldots\}$ must be bounded. Hence, there is a a constant sequence $(x^{\alpha})_n$ converging to π in $\overline{\Omega}_1\underline{S}$, that is $\underline{S} \models \pi = x^{\alpha}$, whence $\pi \in \Omega_1\underline{S}$.

LEMMA 3.2. Let $\pi \in \overline{\Omega}_1\underline{S}$. Then, for $A \in \underline{S}$ and $a \in A$,
a) $\pi_A(a^s) = (\pi_A(a))^s$ for all positive integers s, and
b) $\pi_A(a^{\omega}) = a^{\omega}$.

PROOF. (a) Just note that $b \longmapsto b^s$ defines an endomorphism of any cyclic semigroup.
(b) This follows from (a) noting that there is an integer s such that $b^{\omega} = b^s$ for all $b \in A$.

Let $\underline{G}$ denote the class of all finite groups.

PROPOSITION 3.3 Let $\pi \in \overline{\Omega}_1\underline{S}$. Then, either $\pi \in \Omega_1\underline{S}$, or $\pi(x) = \pi(x^{\omega}x)$ so that π is completely determined by its restriction to $\underline{G}$.

PROOF. Suppose $\pi \notin \Omega_1\underline{S}$. Then, by Lemma 3.1, $\pi_A(a) \in K_a$ for all $A \in \underline{S}$ and $a \in A$. Thus, $\pi_A(a) = \pi_A(a)a^{\omega}$ since a^{ω} is the neutral element of K_a, whence $\pi_A(a) = \pi_A(a)\pi_A(a^{\omega}) = \pi_A(a^{\omega}a)$ applying Lemma 3.2 twice. Hence $\pi(x) = \pi(x^{\omega}x)$ as claimed.

We now study the unary implicit operations on $\underline{G}$. Let $\pi \in \overline{\Omega}_1\underline{G}$. Then, for $A \in \underline{G}$ and $a \in A$, $\pi_A(a) = a^{\alpha(n)}$ for some function $\alpha : \mathbb{N} \longrightarrow \mathbb{N}_0$ where n=ord a is the order of a (since $\pi_A(a) = \pi_{<a>}(a)$). The following Lemma gives the arithmetic conditions on such a function α which insure that the formula $\pi_A(a) = a^{\alpha(n)}$ defines an implicit operation on $\underline{G}$. We write $m|n$ in case m divides n . We denote by $\mathbb{N}$ the set of all positive integers and we let $\mathbb{N}_0 = \mathbb{N} \cup \{0\}$.

LEMMA 3.4. The following are equivalent for a function $\alpha : \mathbb{N} \longrightarrow \mathbb{N}_0$.
i) α defines an implicit operation on $\underline{G}$.
ii) $d|n$ implies $d|\alpha(n) - \alpha(d)$.

PROOF. (i) means, for every $A,B \in \underline{G}$, every homomorphism $h : A \to B$, and every $a \in A$, $h(\pi_A(a)) = \pi_B(h(a))$, i.e.,

$h(a^{\alpha(n)}) = (h(a))^{\alpha(d)}$ where n=ord a and d = ord h(a). Since h is a homomorphism, $d|n$, whence $(h(a))^{\alpha(n)} = (h(a))^{\alpha(d)}$ if and only if $d| \alpha(n)-\alpha(d)$. This proves (ii) $\Rightarrow$ (i). For the converse, just use the above cyclic groups A and B of orders n and d and generators a and b, respectively, and the homomorphism $h : A \to B$ sending a to b.

Let $\mathbb{P}$ denote the set of all primes.

THEOREM 3.5. Let $\lambda_i : \mathbb{P} \to \mathbb{N}$ $(i=0,1,\ldots)$ be given functions and define $\alpha(p^k) = \sum_{i=0}^{k-1} \lambda_i(p)p^i$. Then, there is one and only one extension of α to $\mathbb{N}$ (with $\alpha(n)$ defined up to congruence modulo n) such that α defines an implicit operation on $\underline{G}$. Moreover, every $\pi \in \overline{\Omega}_1\underline{G}$ can be obtained in this way.

PROOF. Let $\pi \in \overline{\Omega}_1\underline{G}$ and let $\alpha : \mathbb{N} \to \mathbb{N}_0$ be any function defining π. Then, by Lemma 3.4, for each $p \in \mathbb{P}$, $p \mid \alpha(p^2) - \alpha(p)$ and so we may assume $\alpha(p^2) = \alpha(p) + \lambda_1(p)p$ for some $\lambda_1(p) \in \mathbb{N}_0$ (since adding any multiple of p^2 to $\alpha(p^2)$ does not change π). An easy induction gives that we may assume $\alpha(p^n) = \alpha(p) + \lambda_1(p)p + \ldots + \lambda_{n-1}(p)p^{n-1}$ with $\lambda_i(p) \in \mathbb{N}_0$ $(i=1,\ldots,n-1)$ independent of n. Finally, take $\lambda_0(p) = \alpha(p)$.

Next we show there is a unique extension $\beta : \mathbb{N} \to \mathbb{N}_0$ of any α defined by a sequence $(\lambda_i)_i$ of functions on prime powers as in the statement of the Theorem so that β defines an implicit operation on $\underline{G}$ (where uniqueness of $\beta(n)$ is again up to congruence modulo n). Indeed, suppose β is such an extension. Let $n \in \mathbb{N}$ and let

$n = p_1^{k_1} \ldots p_k^{k_r}$ be a factorization of n into powers of distinct primes.

Then, by Lemma 3.4, $\beta(n)$ is a solution to the system of congruences

$x \equiv \alpha(p_i^{k_i}) \pmod{p_i^{k_i}}$ $(i=1,\ldots,r)$. By the Chinese Remainder Theorem, this solution always exists and is unique modulo n. This establishes uniqueness and gives a way to define β to establish existence. Thus, it remains to show that β, defined in this way, defines in turn an implicit operation on $\underline{G}$. Now, if $d \mid n$, then

$d = p_1^{\ell_1} \ldots p_r^{\ell_r}$ for some $\ell_i \leq k_i$ $(i=1,\ldots,r)$. Hence

$\beta(n) \equiv \alpha(p_i^{k_i})$ and $\beta(d) \equiv \alpha(p_i^{\ell_i})$ both mod $p_i^{\ell_i}$. Since

$\alpha(p_i^{k_i}) \equiv \alpha(p_i^{\ell_i}) \pmod{p_i^{\ell_i}}$ by the definition of $\alpha(p_i^{k_i})$, it

follows that $p_i^{\ell_i} \mid \beta(n) - \beta(d)$ for $i=1,\ldots,r$ and so $d \mid \beta(n) - \beta(d)$.

By Lemma 3.4, β defines an implicit operation on $\underline{G}$.

COROLLARY 3.6. Every pseudovariety of abelian groups is of the form $[\![\pi=1, xy=yx]\!]$ for some $\pi \in \overline{\Omega}_1\underline{S}$.

PROOF. Let $\underline{V}$ be any pseudovariety of abelian groups. Define $\lambda_i : \mathbb{P} \to \mathbb{N}_0$ by $\lambda_i(p) = 0$ if $\mathbb{Z}_{p^{i+1}} \in \underline{V}$ and $\lambda_i(p) = 1$ otherwise. Let α be as in Theorem 3.5 and $\pi \in \overline{\Omega}_1\underline{S}$ be the implicit operation determined by α. We claim $\underline{V} = [\![\pi=1, xy=yx]\!]$.

Let $A \in \underline{V}$ and $a \in A$ be an element of order n. Let $n = p_1^{k_1} \cdots p_r^{k_r}$ be a factorization of n into products of distinct primes. Then $\alpha(n) \equiv \alpha(p_i^{k_i}) \pmod{p_i^{k_i}}$ $(i=1,\ldots,r)$ and so $\alpha(n) \equiv 0 \pmod{n}$ since $\alpha(p_i^{k_i}) = \sum_{j=0}^{k_i-1} \lambda_j(p)p^j = 0$ as $\mathbb{Z}_{p_i^{k_i}} \in \underline{V}$. Hence,

$\pi_A(a) = a^{\alpha(n)} = 1$. Whence $\underline{V} \models \pi = 1$.

Conversely, let $A \in [\![\pi=1,\ xy=yx]\!]$. Then A is an abelian group since, if $a \in A$, then $\pi_A(a) = a^s$ por some $s \geqq 1$ and a^s is an identity element in A because $A \models \pi = 1$. Thus, A is isomorphic to a direct product $\mathbb{Z}_{p_1^{k_1}} \times \ldots \times \mathbb{Z}_{p_s^{k_s}}$ of cyclic groups (where the p_i are not necessarily distinct primes). If $A \notin \underline{V}$, then some $\mathbb{Z}_{p_i^{k_i}} \notin \underline{V}$ and so, if $\ell = \min\{ m : \mathbb{Z}_{p_i^m} \notin \underline{V} \}$ and $a \in A$ is an element of order p_i^{ℓ}, then $\pi_A(a) = a^{\alpha(p_i^{\ell})} = a^{p_i^{\ell-1}} \neq 1_A$, whence $A \not\models \pi = 1$.

Hence $A \in \underline{V}$.

Using Corollary 3.6, Almeida [2] has observed that there are finite sets Σ of pseudoidentities such that there is no algorithm to decide whether a given finite semigroup S lies in the pseudovariety $[\![\Sigma]\!]$.

COROLLARY 3.7. $|\overline{\Omega}_1 \underline{S}| = 2^{\aleph_0}$.

PROOF. This follows easily from Corollary 3.6 since there are that many pseudovarieties of abelian groups.

4. SOME UNUSUAL BINARY AND TERNARY OPERATIONS

In this section we use some classical results of the combinatorial theory of words to produce, for the values 2 and 3 of n, n-ary implicit operations which are not composites of (n-1)-ary implicit operations and explicit operations.

THEOREM 4.1. Let $\underline{LJ}_1$ denote the class of all finite semigroups S all of whose submonoids eSe $(e^2=e \in S)$ are semilattices. Then, for every pseudovariety $\underline{V}$ containing $\underline{LJ}_1$, there exist binary implicit operations on $\underline{V}$ which are not finite composites of explicit and unary implicit operations.

PROOF. Let $A = \{ a,b \}$ be a two-letter alphabet and consider the endomorphism μ of the free semigroup A^+ on A defined by $\mu(a) = ab$, $\mu(b) = ba$. Consider the sequence $(w_n)_n$ of the words $w_n = \mu^n(a)$ obtained by iteration of μ on a. This sequence of words was first studied by Thuë and Morse and it is known to consist of cube-free words, i.e., no w_n has a factor of the form u^3 with $u \in A^+$ (see Lothaire [6, Chapter 2] for details).

Since $\overline{\Omega}_2\underline{V}$ is compact by Theorem 2.2, $(w_n)_n$ admits a convergent subsequence. We show no subsequence $(w_{\varphi(n)})_n$ can converge in $\overline{\Omega}_2\underline{S}$ to an implicit operation which is a finite composite of explicit and unary implicit operations. Suppose, on the contrary, that $\lim_{n\to\infty} w_{\varphi(n)} = \pi$ where $\pi(a,b)$ is such an implicit operation. Notice that π cannot be an explicit operation.

Then $\pi = \lim_{n\to\infty} v_n$ where $v_n \in A^+$ is given, for all n, by the same finite expression in a,b using the operations $(x,y) \mapsto xy$ and operations of the form $x \mapsto x^{\theta(n)}$, where θ is a given strictly increasing function from $\mathbb{N}$ into itself (non-strictly increasing θ leads to explicit $x \mapsto x^{\theta(n)}$). In particular, if $(v_n)_n$ is not constant (which is the case since $\pi \notin \Omega_2\underline{S}$), v_n has a factor u^3 for all $n \geq 3$ where $u \in A^+$ is independent of n.

Here, we recall that there exists a finite semigroup $S_k \in \underline{LJ}_1 \subseteq \underline{V}$ such that an identity v=w holds in S_k if and only if the words v and w have the same initial and terminal segments of length k-1 and the same factors of length k (cf. Almeida [1]) . Let $k=3|u|$ and let $\ell=|S_k|$. Then, by definition of convergence of sequences in $\overline{\Omega}_2\underline{V}$, there exists $n_0 \geq 3$ such that $S \models w_{\varphi(n)} = \pi = v_n$ for all $n \geq n_0$ and all $S \in \underline{V}$ with $|S| \leq \ell$. In particular, $S_k \models w_{\varphi(n)} = v_n$ so that u^3 is a factor of $w_{\varphi(n)}$, which is impossible since $w_{\varphi(n)}$ is cube-free.

Hence, no accumulation point of the sequence $(w_n)_n$ in $\overline{\Omega}_2\underline{V}$ can be a finite composite of explicit and unary implicit operations.

THEOREM 4.2. For each pseudovariety $\underline{V}$ containing $\underline{LJ}_1$ there exist ternary implicit operations on $\underline{V}$ which are not finite composites of explicit and binary implicit operations.

PROOF. Let $A = \{a,b,c\}$ be a three-letter alphabet. The proof proceeds along the same lines as the proof of Theorem 4.1 working with any infinite sequence of square-free words in A^+ (see also Lothaire [6, Chapter 2] for the existence of such sequences). The only other ingredient is the observation that there are no square-free words of length 4 on a two-letter alphabet. We leave the details to the reader.

In view of Theorems 4.1 and 4.2, it is natural to ask whether in general there are n-ary implicit operations on $\underline{V} \supseteq \underline{LJ}_1$ which are not composites of explicit and (n-1)-ary implicit operations. If the same type of argument as the one used above is going to be applied, one is apparently still lacking theorems on avoidable regularities in words on n-letter alphabets.

5. FINITE SEMIGROUPS IN WHICH IDEMPOTENTS ARE CENTRAL

In spite of the apparent chaos of implicit operations discovered in the previons section, we proceed to clarify their structure. Here, we restrict our attention to the pseudovariety $\underline{ZE} = [\![x^\omega y = yx^\omega]\!]$ of all finite semigroups in which all idempotents are central.

Let $\pi \in \overline{\Omega}_n\underline{ZE}$. We say that π has the <u>kernel property</u> if, for every $S \in \underline{ZE}$ and $s_1,\ldots s_n \in S$, $\pi_S(s_1,\ldots,s_n)$ belongs to the minimal ideal of the subsemigroup of S generated by $s_1,\ldots,s_n$.

LEMMA 5.1. $\underline{ZE} \models (xy)^\omega = x^\omega y^\omega$.

PROOF. Let $S \in \underline{ZE}$ and let n be such that $S \models x^\omega = x^n$. Then

$$S \models (xy)^n = xy(xy)^{n-1}(xy)^n(xy)^n$$
$$= x(xy)^n y(xy)^{x-1}(xy)^n \qquad \text{since idempotents are central}$$
$$= x^2 u_2 (xy)^n \qquad \text{where } u_2 = y(xy)^{n-1}y(xy)^{n-1}$$
$$= x^k u_k (xy)^n \qquad \text{for some } u_k \text{, by induction on k}$$
$$= x^n x^n u_n (xy)^n \qquad \text{since } S \models x^\omega = x^n$$
$$= x^n (xy)^n \qquad \text{by the above.}$$

Analogously, $S \models (xy)^n = (xy)^n y^n$. Furthermore,

$$S \models x^n y^n = xx^{n-1}\, y^n\, x^n\, y^n$$
$$= xyy^{n-1}\, x^{n-1}\, x^n\, y^n$$
$$= xyxx^{n-1}\, y^{n-1}\, x^{n-1}\, x^n\, y^n$$
$$= (xy)^2 (y^{n-1}x^{n-1})^2\, x^n\, y^n$$
$$= (xy)^k (y^{n-1}x^{n-1})^k\, x^n\, y^n \qquad \text{by induction on k}$$
$$= (xy)^n (xy)^n (y^{n-1}x^{n-1})^n\, x^n\, y^n \qquad \text{since } S \models x^\omega = x^n$$
$$= (xy)^n\, x^n\, y^n \qquad \text{by the above.}$$

Hence $S \models x^n\, y^n = (xy)^n\, x^n\, y^n = (xy)^n\, y^n = (xy)^n$.

PROPOSITION 5.2. Let $\pi \in \overline{\Omega}_n \underline{ZE}$. Then π has the kernel property if and only if $\pi(x_1,\dots x_n) = \rho(ex_1,\dots,ex_n)$ where $e=(x_1\dots x_n)^\omega$ for some $\rho \in \overline{\Omega}_n \underline{G}$.

PROOF: Suppose π has the kernel property and let ρ be the restriction of π to $\underline{G}$. Then $\pi(x_1,\dots,x_n) = \pi(x_1,\dots,x_n)e$ since, for each $S \in \underline{ZE}$ and each $s_1,\dots,s_n \in S$, $f = (s_1\dots s_n)^\omega$ is the neutral element of the kernel of the subsemigroup generated by $s_1,\dots,s_n$ by Lemma 5.1. On the other hand, since idempotents are central, the mapping $s \mapsto sf$ is an endomorphism of S. Since π is an implicit operation on $\underline{ZE}$, it follows that $\pi(x_1,\dots,x_n)e = \pi(ex_1,\dots,ex_n) = \rho(ex_1,\dots,ex_n)$. Hence, π is of the required form. The converse is obvious.

LEMMA 5.3. Let $S \in \underline{ZE}$ and let $n=|S|$. Then, for each word w with a number $|w|_x$ of occurrences of a variable x at least n, $S \models w = x^\omega w$.

PROOF. Let $s \in S$ and $t_0,t_1,\dots,t_n \in S^1$. Let $a_k = t_0 s t_1 s \dots t_k s$ $(k=0,\dots,n-1)$. If the a_k $(k=0,\dots,n-1)$ are all distinct, then $a_k = s^\omega$ for some k and so $a_n t_n = s^\omega a_n t_n$. Otherwise, let $k < \ell$ with $a_k = a_\ell$. Then

$$\begin{aligned} a_k = a_\ell &= a_k(t_{k+1}s\dots t_\ell s) \\ &= a_k(t_{k+1}s\dots t_\ell s)^\omega && \text{by the preceding line} \\ &= a_k(t_{k+1}\, s\dots t_\ell s)^\omega s^\omega && \text{by Lemma 5.1} \\ &= s^\omega\, a_k && \text{by the above .} \end{aligned}$$

Hence $a_n t_n = s^\omega a_n t_n$.

We are now ready for the main result of this section .

THEOREM 5.4. Every implicit operation on $\underline{ZE}$ is of the form

$$w_0\, \rho_1(ey_1,\dots,ey_n)\, w_1 \dots \rho_k(ey_1,\dots,ey_n)\, w_k$$

where each w_i is a word not involving the variables $y_1,\dots,y_n$ (with w_0,w_k possibly empty), $e = (y_1\dots y_n)^\omega$, and each $\rho_i \in \overline{\Omega}_n \underline{G}$.

PROOF. Let $\pi \in \overline{\Omega}_m \underline{ZE}$. By Theorem 2.2, there is a sequence $(v_k)_k$ of words on an m-letter alphabet $\{x_1,\dots,x_m\}$ such that $\lim_{k\to\infty} v_k = \pi$.

Let $J = \{j : 1 \leq j \leq m, \{|v_k|_{x_j} : k = 1,2,\dots\}$ is unbounded $\}$.

Then, by considering a subsequence of $(v_k)_k$ if necessary, we may assume that the word u obtained from v_k by removing all x_j with $j \in J$ is the same for all k and $|v_k|_{x_j} \geq k$ for all $j \in J$. Write

$v_k = w_0 v_{k1} w_1 \ldots v_{kr} w_r$ where $w_0 w_1 \ldots w_r = u$.

Given $\ell \in \mathbb{N}$, there exists k_ℓ such that $S \models \pi = v_k$ for all $k \geq k_\ell$ and all $S \in \underline{ZE}$ with $|S| \leq \ell$. In particular, for $k \geq \max\{\ell, k_\ell\}$, $S \models \pi = w_k = x_j^\omega w_k$ $(j \in J)$ by Lemma 5.3.

Let $\sigma_{ki} = v_{ki} \prod_{j \in J} x_j^\omega$ $(i=1,\ldots,r)$ and let $\sigma_k = w_0 \sigma_{k1} w_1 \ldots \sigma_{kr} w_r$. By the above, $\lim_{k\to\infty} \sigma_k = \pi$. Since $\overline{\Omega}_m \underline{ZE}$ is compact, we may assume each sequence $(\sigma_{ki})_k$ converges, say to π_i . Clearly each σ_{ki} has the kernel property and, therefore, so does each π_i . To complete the proof, it suffices to quote Proposition 5.2.

COROLLARY 5.5 (Almeida and Reilly [3]) Every pseudovariety $\underline{V} \subseteq \underline{N}$ is of the form [[$x^\omega = 0, \Sigma$]] for some set Σ of identities.

PROOF. Since $\underline{V} \models x^\omega = 0$, we see that every implicit operation on $\underline{V}$ with the kernel property is constant with value 0 on each $S \in \underline{V}$. Thus, if $\underline{V}$ = [[$x^\omega = 0, \Sigma$]] where Σ is a set of pseudoidentities, then each element of Σ may be replaced by an identity of one of the forms v=0 or v=w in view of Theorem 5.4. Hence, we may assume Σ is a set of identities, as claimed.

For the pseudovariety $\underline{Com}$ of all finite commutative semigroups, we can give a complete description of the implicit operations on $\underline{Com}$.

THEOREM 5.6. Every n-ary implicit operation on $\underline{Com}$ is a product of unary implicit operations (these being viewed as n-ary operations depending on only one variable).

PROOF. Let $\pi \in \overline{\Omega}_n \underline{Com}$. Then, by Theorem 2.2, there is a sequence $(w_k)_k$ of words in $\{x_1,\ldots,x_n\}^+$ such that $\lim_{k\to\infty} w_k = \pi$. Since we are working with commutative semigroups, we may take $w_k = x_1^{\alpha_{k1}} \ldots x_n^{\alpha_{kn}}$ for k=1,2,... . Since $\overline{\Omega}_n \underline{Com}$ is compact, we may assume each sequence $(x_i^{\alpha_{ki}})_k$ converges, say $\pi_i = \lim_{k\to\infty} x_i^{\alpha_{ki}}$. Then $\pi_i \in \overline{\Omega}_1 \underline{Com}$ (i=1,...,n) and $\pi = \pi_1 \ldots \pi_n$.

Theorems 3.5 and 5.6 are, so far, rare complete characterizations of certain sets of implicit operations. In general, it should be more feasible to obtain results like Theorem 5.4 in which implicit operations on a pseudovariety are described modulo the knowledge of implicit operations of some special kind. In this direction, we propose the following conjecture : every implicit operation on $\underline{S}$ is a finite composite of explicit operations and implicit operations which assume only regular values.

REFERENCES

1. Almeida, J., 'Semidirect products of pseudovarieties from the universal algebraist's point of view', in preparation.

2. __________ , 'Pseudovarieties of semigroups', to appear in the *Proceedings of the 1st Meeting of Portuguese Algebraists*, Lisbon 1986 (in portuguese).

3. Almeida, J. and N. R. Reilly, 'Generalized varieties of commutative and nilpotent semigroups', *Semigroup Forum* **30** (1985) 77-98.

4. Eilenberg, S., *Automata, Languages and Machines*, Vol. B, Academic Press, New York, 1976.

5. Lallement, G., *Semigroups and Combinatorial Applications*, Wiley Interscience, New York, 1979.

6. Lothaire, M., *Combinatorics on Words*, Addison-Wesley, Reading, Mass., 1983.

7. Pin, J.-E., *Variétés de Langages Formels*, Masson, Paris, 1984.

8. Reiterman, J., 'The Birkhoff theorem for finite algebras', *Algebra Universalis* **14** (1982) 1-10 .

FINITE IDEMPOTENT-COMMUTING SEMIGROUPS

C.J. Ash,
Department of Mathematics,
Monash University,
Clayton 3168,
Australia.

ABSTRACT

We discuss the result that every finite semigroup in which idempotents commute is a homomorphic image of a subsemigroup of some finite inverse semigroup. The full proof of the general result is to appear elsewhere. In this paper we describe in detail the special case where the relation $\mathcal{J}$ is trivial. This contains most of the features of the general case and we outline what modifications are needed for this.

INTRODUCTION

We discuss the result that every finite semigroup with commuting idempotents is a homomorphic image of a subsemigroup of some finite inverse semigroup. Apart from its intrinsic algebraic interest, this result is also relevant to the study of automata and pseudovarieties, and our argument makes an elementary use of the notion of an injective automaton. We therefore begin with a brief discussion, without proofs, of automata and pseudovarieties.

We next describe in detail an argument which applies in the special case where the relation $\mathcal{J}$ is trivial. We comment here that the method of proof yields a characterization of the languages recognized in this case, and a collection of generators for the pseudovariety.

We proceed to outline how this argument can be modified to suit the general case. This argument is given fully in [1], and answers a question mentioned in [2], [3], [4] and [5].

0. PRELIMINARIES

We work with a fixed, unspecified, finite <u>alphabet</u> $A = \{a_1, a_2, \ldots, a_n\}$. The elements of A are called <u>letters</u>. We denote by A^+ the free semigroup on the set A, which we identify with the set of all finite sequences, or <u>words</u>, from A. It is sometimes convenient to use instead the set $A^* = A^+ \cup \{1\}$, where 1 denotes the empty word. Subsets of A^+ (or of A^*) are called <u>languages</u>.

S. M. Goberstein and P. M. Higgins (eds.), Semigroups and Their Applications, 13–23.

If K is a class of semigroups, then $H(K)$, $S(K)$ and $P_f(K)$ denote, respectively, the classes of all homomorphic images, subsemigroups and finite direct products of members of K. A class K of finite semigroups is said to be a pseudovariety if $H(K) \subseteq K$, $S(K) \subseteq K$ and $P_f(K) \subseteq K$ or, equivalently, if $HSP_f(K) = K$. Thus, every class K of finite semigroups is included in a smallest pseudovariety, namely $HSP_f(K)$, also called the pseudovariety generated by K.

We say that one semigroup, S, divides another, S', if there is a subsemigroup T of S' and a surjective homomorphism from T onto S. We denote by $\mathcal{PT}(X)$ and $\mathcal{I}(X)$, respectively, the semigroups of all partial functions and of all partial one-one functions from the set X into itself. We use the abbreviation "i.c." for "idempotent-commuting". Thus an i.c. semigroup is one for which $ef = fe$ whenever e and f are idempotents.

1. LANGUAGES AND SEMIGROUPS

We say that a language L is recognized by a finite semigroup S if there is a homomorphism $\phi : A^+ \to S$ and a subset F of S for which $L = F\phi^{-1}$.

Thus, for different choices of ϕ and F, a semigroup S recognizes each of some finite set, $\mathcal{L}(S)$, of languages. More generally, for a class K of finite semigroups, let $\mathcal{L}(K)$ denote the set of all those languages which are recognized by at least one member of K.

The following is easily seen from the definitions.

Lemma 1.1

(i) $\mathcal{L}(HS(K)) = \mathcal{L}(K)$.

(ii) $\mathcal{L}(K)$ is closed under complements.

(iii) If $K = P_f(K)$, then $\mathcal{L}(K)$ is closed under unions and intersections.

□

Thus, if a class K of finite semigroups is of interest, then so is the class $HS(K)$ of all semigroups which divide some member of K, since to show that a language is recognized by a member of K we need only show that it is recognized by a member of $HS(K)$. If also $K = P_f(K)$ then $\mathcal{L}(K)$ is a Boolean algebra and $HS(K) = HSP_f(K)$ is the pseudovariety generated by K. In the case where K is the class of all finite inverse semigroups, we shall see that this pseudovariety consists of exactly the finite i.c. semigroups.

We may quickly dispose of the "easy" direction of this result.

Lemma 1.2

If a semigroup S divides a finite inverse semigroup, the S is i.c.

Proof

Let I be a finite inverse semigroup, T be a subsemigroup of I and $\theta : T \to S$ be a surjective homomorphism. Let e, f be idempotents in S. Then $\{x \in T : x\theta = e\}$ is a finite subsemigroup of T and so contains at least one idempotent, e'. Similarly let f' be an idempotent of T for which $f'\theta = f$. Since e' and f' are elements of the inverse semigroup I, we have $e'f' = f'e'$, and so

$$ef = (e'\phi)(f'\phi) = (e'f')\phi = (f'e')\phi = (f'\phi)(e'\phi) = fe. \quad \square$$

To show that every finite i.c. semigroup is in the pseudovariety, we need the converse of this lemma. Our arguments for this are most easily expressed in terms of injective automata.

2. AUTOMATA

For our purposes we may define an automaton to consist of a non-empty finite set X, an element i of X, a subset F of X and a homomorphism $\phi : A^+ \to \mathcal{PT}(X)$. The automaton is said to accept a word $w \in A^+$ if $i(w\phi)$ is defined and belongs to the set F. The set of all $w \in A^+$ which are accepted is called the language recognized by the automaton.

So such a language, L, is recognized in the previous sense by the semigroup $\mathcal{PT}(X)$. Conversely, if a language L is recognized by a semigroup S, then we may construct an automaton recognizing L by using the right regular representation of S^1, thus proving the following.

Lemma 2.1

A language is recognized by some finite semigroup iff it is recognized by some automaton. $\square$

Corresponding to inverse semigroups we naturally define an injective automaton to be an automaton, as before, except that ϕ is a homomorphism from A^+ to $\mathcal{I}(X)$. Here, however, the languages recognized are not the same as for inverse semigroups, although as a matter of interest they are related as follows.

Lemma 2.2

(i) A language is recognized by some finite inverse semigroup iff it is a Boolean combination of languages recognized by injective automata.

(ii) A language is recognized by an injective automaton iff it is $F\phi^{-1}$ for some finite inverse semigroup I, some homomorphism $\phi : A^+ \to I$ and some subset F of I satisfying the following condition: whenever $s \in F$ and $s \leq t$ in the natural ordering of the inverse semigroup, then $t \in F$.

Our use of injective automata, however, depends only on the following lemma, essentially (i) above.

Lemma 2.3

Let S be a finite semigroup and let $\psi : A^+ \to S$ be a surjective homomorphism. Then the following are equivalent.

(1) There is a finite inverse semigroup I for which S divides I.

(2) There is a number K such that, whenever $w, w' \in A^+$ and $w\psi \neq w'\psi$, then there is an injective automaton (X,i,ϕ,F) for which $|X| \leq K$ and for which the language recognized contains one, but not both, of w and w'. □

Comment

Under the conditions of (2), we shall say that the automaton separates w and w'. Clearly we may then assume that $F = \{t\}$ for some $t \in X$.

For an alphabet $A = \{a_1,\ldots,a_n\}$, and an automaton (X,i,ϕ,F), the homomorphism ϕ is determined by the n partial functions $a_1\phi,\ldots,a_n\phi$. So the automaton can be briefly described by its graph, consisting of the set X of vertices and, for each j, a directed edge labelled by a_j from x to y for each pair (x,y) of vertices such that $x(a_j\phi) = y$. The vertex i is labelled by the letter i and the vertices in F by the letter t.

3. $\mathcal{J}$-TRIVIAL I.C. SEMIGROUPS

We consider the special case of a finite i.c. semigroup, S, in which the relation $\mathcal{J}$ is trivial. Our treatment of this case contains most of the salient features of the general case.

To use Lemma 2.3, we fix a surjective homomorphism $\psi : A^+ \to S$, for sufficiently large A, and consider $w, w' \in A^+$ for which $w\psi \neq w'\psi$. We may always find a sufficiently large automaton which separates w and w' in the sense of the Lemma. However, we must find one whose size is bounded independently of the lengths of w, w'.

As we shall see, the case where both $w\psi$ and $w'\psi$ are idempotent is very easy. Led by this we first ask, for a single $w \in A^+$, which factors v of w are such that $v\phi$ is idempotent. The following lemmas clarify this question, for finite $\mathcal{J}$-trivial i.c. semigroups.

Lemma 3.1

If $r, s, t \in S^1$ and rst is idempotent, then $(rst)s = s(rst) = rst$.

Proof

Since rstrst = rst, we have rst $\mathcal{R}$ rstr and so rst = rstr, since $\mathcal{J}$ is trivial. Likewise, rst $\mathcal{R}$ rstrs, so rst = rstrs. Now substituting gives rst = rsts. Similarly, trst = rst and strst = rst, and so srst = rst. □

Lemma 3.2

If $r,s,t \in S^1$ and rs, st are idempotents then rst is idempotent.

Proof

Certainly each product of idempotents is idempotent, since idempotents commute. So $(rs)(st) = rs^2t$ is idempotent. But by Lemma 3.1, $rs^2 = rs$, because $rs^2 = (rs\,1)s = rs\,1 = rs$. Hence $rst = rs^2t$ is idempotent.

Lemma 3.2 shows that the factors v of $w \in A^+$ for which $v\psi$ is idempotent behave in a very simple way.

For each $w \in A^+$, let $w = b_1b_2 \ldots b_r$ where each $b_i \in A$. Say that an interval $[i,j] = \{i,i+1,i+2,\ldots,j\}$, where $1 \leq i \leq j \leq r$, is good if $(b_ib_{i+1} \ldots b_j)\psi$ is idempotent in S. Then by Lemma 3.2, for every two good intervals which are adjacent or overlapping, their union is also a good interval. Hence, any two different good intervals which are maximal under set inclusion are disjoint and separated by at least one integer.

Let $[i_1,j_1], [i_2,j_2],\ldots,[i_m,j_m]$ be the good maximal intervals in increasing order. We define v_p to be the product of the b_i for $i_p \leq i \leq j_p$ and u_p to be the product of the b_i for $j_p < i < i_{p+1}$, with the conventions that $j_0 = 0$ and $i_{m+1} = r+1$.

The corresponding factorization, $w = u_0v_1u_1 \ldots v_mu_m$ will be called the canonical factorization of w (with respect to ψ).

Comment

This factorization is thus uniquely determined by ψ and S. In fact, for our purposes it would be sufficient to take a factorization in which each v_i is the product of a good interval and the total length of the u_i is minimum. But the additional information given by Lemma 3.2 seems well worth observing.

Theorem 3.3

If $w = u_0v_1u_1 \ldots v_mu_m$ is the canonical factorization of w, then:

(i) Each u_i, except perhaps u_0 or u_m, is non-empty.

(ii) Each $v_i\psi$ is idempotent in S.

(iii) Neither the last letter of u_{i-1} nor the first letter of u_i appears in the word v_i.

Proof

(i) and (ii) are clear from the previous discussion. For (iii), suppose that b is the first letter of u_i. Since v_i corresponds to a maximal good interval, $(v_i b)\psi$ is not idempotent. But if b appears in v_i, then $v_i = xby$ where x,y A^*. So $(v_i b)\psi = (x\psi)(b\psi)(y\psi)(b\psi) = (x\psi)(b\psi)(y\psi) = (xby)\psi = v_i\psi$, using Lemma 3.1, contradiction. Similarly if b is the last letter of u_{i-1}. □

The canonical factorization is splendid for our purpose because of the next theorem. This follows easily from the consequence of Ramsey's Theorem which asserts that for every finite semigroup S there is a number $k(S)$ such that for every sequence $s_1, s_2, \ldots, s_m$ with $m > k(S)$ there exist i, j with $1 \leq i \leq j \leq m$ for which the product $s_i s_{i+1} s_{i+2} \ldots s_j$ is idempotent.

Theorem 3.4

Let $w \in A^+$ and let $w = u_0 v_1 u_1 \ldots v_m u_m$ be its canonical factorization. Then the sum of the lengths of the words u_i is at most $k(S)$.

Proof

Let $u_i = c_{i1} \ldots c_{ir_i}$ where each $c_{ij} \in A$. Consider the sequence

$$c_{01}\psi, c_{01}\psi, \ldots, c_{0r_0}\psi, (v_1 c_{11})\psi, c_{12}\psi, \ldots, (v_2 c_{21})\psi, c_{22}\psi, \ldots,$$

$$\ldots (v_m c_{m1})\psi, c_{m3}\psi, \ldots, c_{mr_m}\psi.$$

If the total length of the u_i exceeds $k(S)$, then so does the length of this sequence. In this case, by Ramsey's Theorem, there is a consecutive subsequence whose product is idempotent. But this gives a good interval containing the subscript of a letter in at least one of the u_i contradicting the definition of the factorization.

Note

In the canonical factorization of w, we may have $m = 0$ and so $w = u_0$. But by Theorem 3.4, this can only happen when the length of w does not exceed $k(S)$.

Definition

For each $w \in A^+$, let $c(w)$ denote the set of all letters from A appearing in w.

We may now see how to separate all pairs $w, w' \in A^+$, for which $w\psi \neq w'\psi$, in an injective automaton having at most K_r elements, where $r = |c(w) \cup c(w')|$. Our procedure is by induction on r and, of course, we take K in Lemma 2.3 to be K_r when $r = |A|$.

Case 1

One of w, w' has at most $k(S)$ elements. If this word is $b_1b_2 \ldots b_s$ where each $b_i \in A$, then we may use the automaton below.

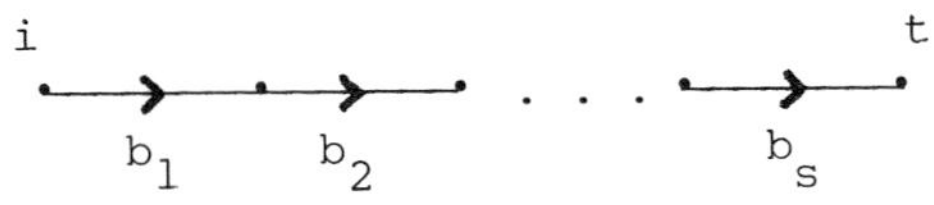

This accepts only the word $b_1b_2 \ldots b_s$ and will suffice since we are at liberty to choose each $K_r \geq k(S)-1$.

Case 2

Each of $w\psi$, $w'\psi$ is idempotent. In this case, Lemma 3.1 shows that if every letter occurring in w' also occurs in w, then $(ww')\psi = w\psi$. Hence, w.l.o.g., $c(w') \not\subseteq c(w)$, and we may use the following automaton.

Here the notation indicates one edge for each letter in $c(w)$. This automaton therefore accepts w but not w'.

Case 3

If neither case 1 nor case 2 applies, then w.l.o.g. w has canonical factorization $u_0v_1u_1 \ldots v_mu_m$ where $m \geq 1$ and at least one u_j is non-empty. Let each $u_j = b_{j1}b_{j2} \ldots b_{jr_i}$ where each $b_{jk} \in A$ and consider first the following automaton.

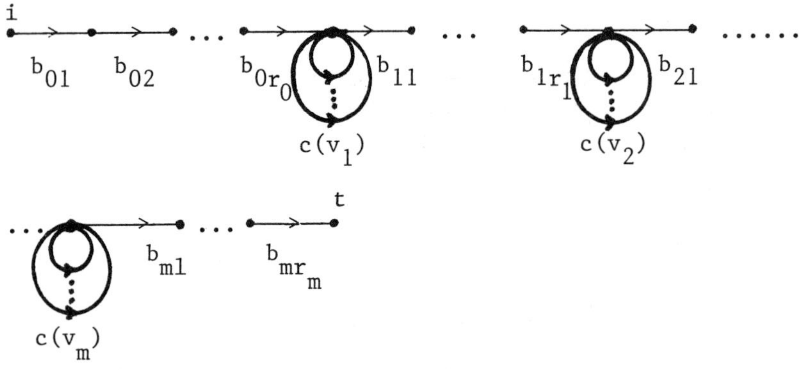

This is injective, by Theorem 3.3(iii), and accepts w. If it does not accept w' then we need go no further since we have already decided that $K_r \geq k(S)-1$. Otherwise, it does accept w' and so

$w' = u_0 v'_1 u_1 v'_2 u_2 \ldots v'_m u_m$ where $c(v'_j) \subsetneq c(v_j)$. This need not be the canonical factorization of w', but certainly since $w\psi \neq w'\psi$ there must be at least one j for which $v_j\psi \neq v'_j\psi$.

By Theorem 3.3(iii), $c(v_j) \not\subseteq c(w)$. Since also $c(v'_j) \subsetneq c(v_j)$ we may apply the induction hypothesis to obtain an injective automaton

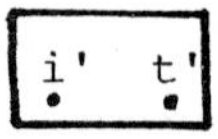

which separates v_j and v'_j and which needs only to involve the letters in $c(v_j)$. Now we may modify the previous automaton and replace the part

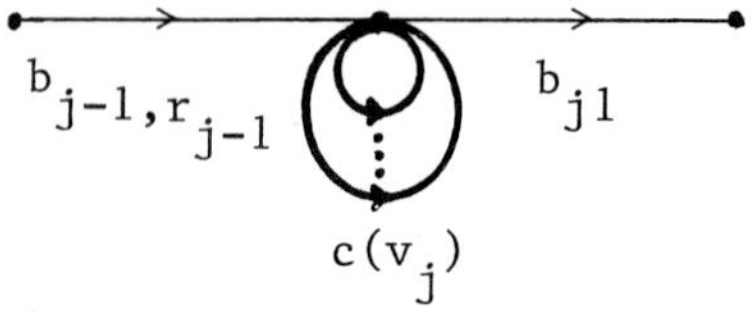

by

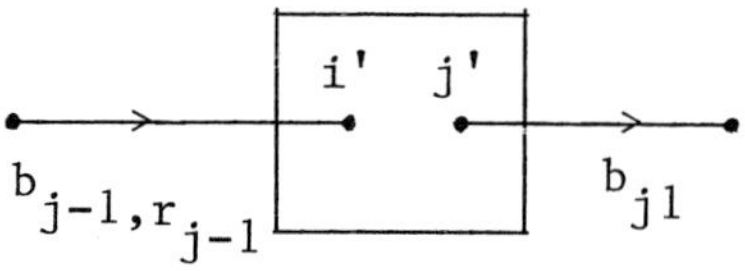

We can do this irrespective of whether $i' = j'$. The result accepts one but not the other of w, w' because $b_{j1} \notin c(v_j)$.

Since we have only added at most $k(S)-2$ elements to the inserted automaton, we need only ensure that $K_{r+1} \geq K_r + k(S) - 2$.

In view of Lemma 2.3 we have therefore proved:

<u>**Theorem 3.5**</u>

If S is finite, i.c. and $\mathcal{J}$-trivial, then S divides some finite inverse semigroup. $\square$

A further consequence of our construction concerns languages. We note that for each of the automata constructed, the languages recognized are of the form $A_0^*\{b_1\}A_1^*\{b_2\}\ldots\{b_k\}A_k^*$ where each $b_i \in A$, each $A_i \subsetneq A$ and for each i, both $b_{i-1} \notin A_i$ and $b_i \notin A_i$. (We allow A_i to be empty, in which case $A_i^* = \{1\}$.)

Now our construction shows that if $w, w' \in A^*$ and $w\psi \neq w'\psi$ then w and w' are separated by one of finitely many languages of this form. This immediately gives the following.

<u>**Theorem 3.6**</u>

The languages recognized by finite i.c. $\mathcal{J}$-trivial semigroups are the Boolean combinations of those of the form $A_0^*\{b_1\}A_1^* \ldots\{b_n\}A_n^*$ where $b_i \in A$, $A_i \subsetneq A$ and $b_{i-1}, b_i \notin A_i$. $\square$

For the same reason we have:

Theorem 3.7

The pseudovariety of all finite i.c. $\mathcal{J}$-trivial semigroups is generated by the semigroups S_n where S_n is the semigroup of all one-one non-decreasing partial functions on $\{1,2,\ldots,n\}$.

4. THE GENERAL CASE

The canonical factorization described in §3 may be generalized for the case where S is finite and i.c. but $\mathcal{J}$ is not necessarily trivial. We give an outline here. Full proofs are given in [1].

Again we fix a surjective homomorphism $\psi : A^+ \to S$. Letting $w = b_1 b_2 \ldots b_r$, we now define a good interval to be an interval $[i,j]$ for which $(b_i b_{i+1} \ldots b_j)\psi$ is a regular element of S.

The appropriate lemma is now:

Lemma 4.1

If $r,s,t \in S^1$ and rs, st are regular, then so is rst. □

This shows that the distinct maximal good intervals are again disjoint and separated by at least one element. We define the canonical factorization $w = u_0 v_1 u_1 \ldots v_m u_m$ as before, in terms of the maximal good intervals.

We may now show, much as for Theorems 3.3 and 3.4:

Theorem 4.2

Let $w = u_0 v_1 u_1 \ldots v_m u_m$ be the canonical factorization of w (w.r.t. ψ and the regular elements of S).

(i) Each u_i, except perhaps u_0 or u_m, is non-empty.

(ii) Each $v_i\psi$ is regular.

(iii) If b is the last letter in u_{i-1}, then $(bv_i)\psi$ is not regular. If b is the first letter in u_i then $(v_i b)\psi$ is not regular.

(iv) The sum of the lengths of the u_i is at most k(S). □

In the place of the one-element automata used in §3 for words v such that $v\psi$ is idempotent, we may use, for $v\psi$ regular, automata obtained from the $\mathcal{R}$-classes of S.

For each $\mathcal{R}$-class, R, of S and each $s \in S$, we define the partial function ρ_s on R by
$$x\rho_s = \begin{cases} xs & \text{if } xs \in R \\ \text{undefined otherwise.} \end{cases}$$

Lemma 4.3

Each ρ_s is one-one. The map $\phi : A^+ \to I(R)$ defined by $w\phi = \rho_{w\psi}$ is a homomorphism. □

Lemma 4.4

If $v, v' \in A^+$, $v\psi$ and $v'\psi$ are regular and $v\psi \neq v'\psi$ then v and v' are separated by an injective automaton obtained from an $\mathcal{R}$-class.

Proof

W.l.o.g. $vv^{-1} \not\leq v'(v')^{-1}$, so we may take the $\mathcal{R}$-class of vv^{-1} with $i = vv^{-1}$, $t = v$. □

The same automaton has a further property.

Lemma 4.5

If $v \in A^+$, $b \in A$, $v\psi$ is regular and $(vb)\psi$ is not, then there is an $\mathcal{R}$-class automaton $(R, i, \phi, \{t\})$ for which $i(v\phi) = t$ and $i((vb)\phi)$ is undefined. □

Similarly, if $v\psi$ is regular and $(av)\psi$ is not then we may construct an injective automaton $(L, i, \phi, \{t\})$ from an $\mathcal{L}$-class in which $i(v\phi) = t$ and for no x is $x((av)\phi) = i$. Taking the direct product of these automata gives the following.

Lemma 4.6

If $v \in A^+$, $a, b \in A$, $v\psi$ is regular but $(av)\psi$ and $(vb)\psi$ are not, then there is an injective automaton $(X, i, \phi, \{t\})$ with $|X| \leq |S|^2$, $i(v\phi) = t$, t not in the domain of $b\phi$ and i not in the range of $a\phi$. □

Using this, we may now imitate our previous argument and show that there is a K, depending only on S, such that for $w, w' \in A^+$ and $w\psi \neq w'\psi$, there is an injective automaton with at most K elements separating w and w'.

Let $w = u_0 v_1 u_1 \ldots v_m u_m$ be the canonical factorization of w. By Lemma 4.6, we may choose for each i an injective automaton for v_i such that the following composite automaton is injective.

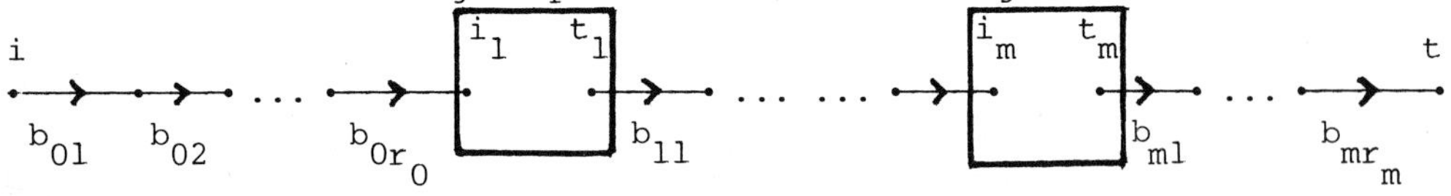

(As before $u_i = b_{i1} \ldots b_{ir_i}$.)

If this automaton does not separate w and w' then $w' = u_0 v'_1 u_1 \ldots v'_m u_m$. Since $w\psi \neq w'\psi$, for some i, $v_i\psi \neq v'_i\psi$. Now by the induction hypothesis explained below there is an injective automaton separating v_i and v'_i. We do not substitute <u>this</u> for the previous automaton corresponding to v_i, but rather the direct product of the two. This will ensure that the resulting composite automaton accepts only one of w, w', as required.

The induction is valid because, except when $w\psi$ is regular (so $w = v_1$), the $\mathcal{J}$-class of $v_i\psi$ is strictly above that of $w\psi$. Thus we can proceed by induction on the sum of the numbers of $\mathcal{J}$-classes above $w\psi$

and $w'\psi$. The extreme cases are where either w or w' has length at most $k(S)$, dealt with as in §3, or both $w\psi$ and $w'\psi$ are regular, which is covered by Lemma 4.4.

In view of Lemma 2.3, this establishes:

Theorem 4.7

Every finite i.c. semigroup divides some finite inverse semigroup. □

Consequently, by Lemma 1.2, we have:

Theorem 4.8

The pseudovariety generated by the finite inverse semigroups consists of exactly the finite i.c. semigroups. □

REFERENCES

[1] Ash, C.J., 'Finite semigroups with commuting idempotents'. J. Aust. Math. Soc. (to appear).

[2] Margolis, S.W. 'Problem 14', Proceedings of the Nebraska conference on semigroups, (Ed. J. Meakin), September 1980, p. 14.

[3] Margolis, S.W. and Pin, J.E. 'Languages and inverse semigroups'. 11th ICALP, Lecture Notes in Computer Science **172** (1984), 337-346.

[4] ———————————— 'Graphs, inverse semigroups and languages'. Proceedings of the 1984 Marquette conference on semigroups. Marquette University (1984), 85-112.

[5] Pin, J.E. Variétés de langages formels. Masson, Paris (1984).

FINITE SEMIGROUPS WHOSE IDEMPOTENTS COMMUTE OR FORM A SUBSEMIGROUP

Jean-Camille Birget and Stuart Margolis
Department of Computer Science
University of Nebraska
Lincoln, Nebraska

and

John Rhodes
Department of Mathematics
University of California
Berkeley, California

INTRODUCTION. We give a new proof that every finite semigroup whose idempotents commute divides a finite inverse semigroup (Ash's theorem), and, more generally, we prove that every finite semigroup whose idempotents form a subsemigroup divides a finite orthodox semigroup.

Our proof considers relational morphisms from a finite semigroup into finite groups, and examines the elements of S that are always related to the identity of these groups. Such elements were first studied by Rhodes and Tilson [R.T.] (under the name of "type II elements"), and later, in a different context, by McAlister [Mc] for arbitrary regular semigroups.

It follows from a theorem by Margolis and Pin [M.P.] that if the idempotents of S (finite) commute then we have: S divides a finite inverse semigroup iff the type II elements of S form a semilattice. In addition, Thérien and Weiss [T.W.] showed that a finite semigroup S divides an orthodox semigroup iff the type II elements form a band.

The main part of our proof consists then in showing that if the idempotents of S form a band (this covers the case where they are a semilattice) then all the type II elements of S are idempotents. Ash's ideas (in [A]) were a good source of inspiration for carrying out the proof. The expansion R(S) (section 2) and its use in defining partial injective actions on states (similar to section 4) had been presented by Birget at the Chico Semigroup Conference.

1) Type II elements of a finite semigroup

Definition. Let S and T be semigroups. A relational morphism from S into T is a subsemigroup Φ of $S \times T$ (direct product) such that for all $s \in S$ there exists $t \in T$ with $(s,t) \in \Phi$ (i.e. every element of S is related to some element of T).

In this section we assume S is a finite semigroup, and we consider

S. M. Goberstein and P. M. Higgins (eds.), Semigroups and Their Applications, 25–35.

all relational morphisms of S into finite groups.

Definition [R.T]. An element s of a finite semigroup S is a type II element iff for every finite group G and every relational morphism ϕ of S into G we have: s is related to the identity 1 of G (i.e.: $(s,1) \in \phi$).

Elementary facts

(1) The set of type II elements of S is a subsemigroup of S (which we will denote by S_{II}).

(2) Every idempotent of S belongs to S_{II}.

(3) If $s \in S_{II}$, and the elements r, x of S satisfy $r\,x\,r = r$ (so r is regular, but x might be non-regular), then rsx and xsr belong to S_{II}. (So S_{II} is closed under "weak conjugation".)

(4) If $s \in S_{II}$, and r is an element of S satisfying $rsr = r$, then r belongs to S_{II}.

Property (4) implies that if $s \in S_{II}$ and is a regular element of S then s is actually regular in S_{II} (s has an inverse in S_{II}).

The proof of these elementary facts is straight forward.

Rhodes and Tilson [R.T.] proved that all the regular elements of S_{II} are contained in the smallest subsemigroup of S satisfying properties (1), (2) and (3) above.

Remark. If in the definition of "type II" the groups are allowed to be infinite then one can easily show that no elements of S (not even the idempotents) are always related to the identity. The groups must at least be torsion.

The importance of type II elements in this context arises from the following theorems:

Theorem 1.1 (Margolis, Pin). Let S be a finite semigroup. Then S divides a finite inverse semigroup iff S_{II} is a semilattice.

The proof of this theorem (in [M.P.2]), uses Tilson's derived category together with a lemma of I. Simon, that describes the free idempotent and commutative category over a graph.

Denis Thérien and Alex Weiss [T.W.] proved a similar result for the free idempotent category over a graph. Using this one proves:

Theorem 1.2 (Thérien, Weiss). Let S be a finite semigroup. Then S divides a finite orthodox semigroup iff S_{II} is a band.

The main part of our paper consists in showing that if the idempotents of S commute (respectively, form a band) then S_{II} is a semilattice (resp. a band).

In fact, all we need to show is:
<u>If S is a finite semigroup whose idempotents form a band then S_{II} is regular</u>.

Indeed we have:

Proposition: <u>If S is a finite semigroup whose idempotents form a band and for which S_{II} is regular, then S_{II} is a band</u>.

If in addition, the idempotents of S commute, then they also commute in S_{II}, hence in this case S_{II} is a semilattice.

<u>Proof</u>: By the results of Rhodes and Tilson (stated after the "Elementary facts"), we know that the regular elements of S_{II} are obtained by taking idempotents and repeatedly applying weak conjugation and multiplication. However, if the idempotents of S form a band then the product of idempotents is an idempotent. Moreover, the weak conjugates xer and rex of an idempotent e (where r x r = r) are idempotents: indeed x e r = x e r x r = x e r x e r x r (since erx is an idempotent); hence = x e r x e r. The case of rex is similar.

In order to prove that every element of S_{II} is regular (if the idempotents of S form a band) we introduce a new expansion.

2) An expansion

Simply stated, an expansion associates to every semigroup S a semigroup <u>Ex</u>(S) such that <u>Ex</u>(S) $\twoheadrightarrow$ S (i.e. S is a homomorphic image of <u>Ex</u>(S)). The full definition of an expansion can be found in [B.R.] but it will not be needed here.

For any semigroup S we define the expansion R(S) to be the semigroup presented by <u>generators and relations</u> as follows:
Generators: the set S.
Relations: the set $\{w = \Pi w \,/\, w \in S^+$, and Πw is <u>regular</u> in S$\}$.

Here we use the following <u>notation</u>:
S^+ is the set of all finite non-empty sequences of elements of S.
If $w = (a_1,\ldots,a_n) \in S^+$ then $\Pi w = a_1 \ldots a_n$ (i.e. Πw is the product in S of all the terms in the sequence w).

The semigroup S is a homomorphic image of R(S) via the map defined on representatives by: $w \to \Pi w$. This is clearly a well defined surmorphism R(S) $\twoheadrightarrow$ S, which we will denote by Π (the product map).

Theorem (Properties of the expansion R(S)).

Let S be any semigroup.

(a) For every $w \in R(S)$ we have: w is regular in R(S) iff Πw is regular in S iff $\Pi w = w$ (i.e. w and Πw are equal in R(S)). So one can say that the regular elements of R(S) and S are "the same".

It follows that if idempotents of S commute (resp. form a band), the same will be true in R(S).

(b) If S is a **finite** semigroup then R(S) is also finite.

Proof: Part (a) of the theorem follows immediately from the defining relations of R(S), and from the fact that homomorphic images (via the product map Π in this case) of regular elements are regular.

For a proof of part (b) see [B.M.R.]. One can use Brown's theorem [Br], or Ash's method (Ramsey's theorem) [A], or techniques of Birget [Bi].

Fact (Representatives (in S^+) of the elements of R(S))

(a) Every **regular** element of R(S) can be identified with a unique regular element of S.

(b) Every **non-regular** element of R(S) is equivalent (via the defining relations of R(S)) to a sequence of the form $w = (n_0, r_1, n_1, \ldots, r_k, n_k)$ where:

each r_i is a **regular** element of S,

each n_i is a (possibly empty) sequence of non-regular elements of S with the property that Πn_i is **non-regular** in S.

(c) For every subsegment x of length >1 of w we have:

Πx is non-regular (i.e. no rule $u \to \Pi u$, with Πu regular can be applied to w).

(d) If the **regular elements** of S **form a subsemigroup** (in particular, this holds if the idempotents of S form a band) then every element of R(S) has a **unique** representative w satisfying properties (a), (b), (c) above.

Proof: Parts (a), (b) and (c) are straight forward. Part (d) is a consequence of the following.

Lemma. Let S be a semigroup whose regular elements form a subsemigroup. Then for all $x, y, z \in S$ we have:

If both xy and yz are regular then xyz is also regular.

Proof of the lemma. See [B.M.R.].

The Fact and the Lemma were first proved by Ash [A] in the case where S is a finite semigroup whose idempotents commute.

3) Proof that S_{Π} contains no non-regular elements if the idempotents of S form a band: Introduction

For every non-regular element s of S we will construct a relational morphism Φ from S into a group G such that $(s,1) \notin \Phi$ (s is not related to the identity element of G). We only deal with finite groups or semigroups.

It will be sufficient to deal with full (finite) symmetric groups, and (finite) direct products of such groups. We will denote the full symmetric group on a set Q by G(Q).

A relational morphism of S into a given group G can be obtained by first picking a subset $G_s \subseteq G$ for each $s \in S$, and then considering the subsemigroup of $S \times G$ generated by $\{(s,g) / s \in S, g \in G_s\}$.

A very useful way of constructing a relational morphism from a semigroup S into the symmetric group G(Q) (for a given set Q) is as follows: (1) Let every element s of S "act" on Q as a partial injective function (i.e. to every element $s \in S$ associate a partial function $f_s: Q \to Q$ which is injective). Here we do not require that $f_{st} = f_s f_t$. (2) Next we extend each chosen partial injection f_s to a (total) permutation p_s on Q. (3) Then we obtain a relational morphism of S into the symmetric group G(Q) by taking the subsemigroup Φ of $S \times G(Q)$ generated by $\{(s,p_s)/s \in S\}$. Notice that now $(s,p) \in \Phi$ iff s can be factored as $s = s_1 \ldots s_k$, and $p = p_{s1} \ldots p_{sk} \in G(Q)$.

What we will need is a little more general than that: we will need a relational morphism of S into a direct product of symmetric groups $G(Q_1) \times \ldots \times G(Q_n)$. The way to do that is similar to the above. (1) First let every element s of S act as a partial injective function $f_{s,i}$ on each Q_i (for $1 \leq i \leq n$). (2) Next extend each $f_{s,i}$ to a total permutation $p_{s,i} \in G(Q_i)$. (It does not matter how $f_{s,i}$ is extended, provided that the result is a permutation.) (3) Finally, we obtain a relation of S into $G(Q_1) \times \ldots \times G(Q_n)$ generated by $\{(s,p_{r,1},\ldots,p_{s,n})/s \in S$, and $p_{s,i} \in G(Q_i)$ for $1 \leq i \leq n\}$.

In this situation, s is related to the identity (that is: $(s,1,\ldots,1) \in \Phi$) iff there exists a factorization of s as $s_1 \ldots s_k$ such that for all i with $1 \leq i \leq k$ we have $p_{s1,i} \ldots p_{sk,i} = 1$

Contrapositively, s does not belong to S_{II} iff for all factorizations of s as $s = s_1 \ldots s_k$ there exists i ($1 \leq i \leq k$) such that $p_{s1,i} \ldots p_{sk,i} \neq 1$.

We shall next construct the state sets Q_i, and then show that if s is non-regular then $s \notin S_{II}$.

4) State sets and actions

In this section we assume that S is a finite semigroup whose idempotents form a band.

We will use the following equivalence relation on the set of regular elements of S :

Definition. Let x, y be regular elements of S. Then $x \sim y$ iff x and y have a common inverse.

Fact 4.0 If S is a semigroup whose idempotents form a band and if x and y are regular elements of S such that $x \sim y$, then x and y have exactly the same inverses.

Fact 4.1. If the idempotents of S form a band then $\sim$ is a congruence relation on the subsemigroup of regular elements of S.

Proof of 4.0 and 4.1: See e.g. [H.].

In the following we will call the $\sim$ congruence classes "blocks". The congruence class of $r(\varepsilon S)$ will be denoted by $[r]$.

To define our states we will combine the expansion R(S) and the congruence $\sim$:

First to every $\sim$ congruence class we associate a distinct symbol. Let $B = \{b_1,\ldots\}$ be the set of such symbols (names of blocks). We will also denote the symbol of $[r]$ by $b(r)$.

Next, to each reduced word $w = (n_0,r_1,n_1,\ldots,r_k,n_k)$ representing an element of R(S) we associate the word

$(n_0,b(r_1),n_1,\ldots,b(r_k),n_k)$, over the alphabet $B \cup N$ (where N is the set of non-regular elements of S). We will denote the set of these words by $R_b(S)$.

Remark. The set $R_b(S)$ can be turned into a semigroup which is a homomorphic image of R(S). It can be presented by the generators $N \cup B$ and the relations $\{b(\Pi w) = w \,/\, w \in S^+,\ \Pi w \text{ regular in } S\}$. However we will not directly use the semigroup structure of $R_b(S)$.

States

For every word $v \in R_b(S)$ we introduce a set of states Q_v.

First we define a start state of Q_v : If $v = (n_0,\ldots)$ with n_0 non-empty, then the start state of Q_v is the empty word 1. If $v = (b(r_1),\ldots)$ (i.e. n_0 is empty) then $b(e_1)$ is the start state, where e_1 is any idempotent with $e_1 \equiv_R r_1$. If the idempotents of S form a band then $b(e_1)$ is unique (by fact 4.2 below).

Let $v = (n_0, b(r_1),n_1,\ldots,b(r_k),n_k)$. Then Q_v consists of the start state and of words of the following form:

First, $(n_0,\ldots,n_{i-1},b(r))$ where $r \equiv_R r_i$, $1 \le i \le k$ (and the segment $(n_0,\ldots,n_{i-1})$ coincides with the corresponding segment of v).

Second, $(n_0,\ldots,(a_1^{\,i-1},\ldots,a_j^{\,i-1}))$ where $1 \le i \le k$, and

$(a_1{}^{i-1},\ldots,a_j{}^{i-1})$ is a prefix of n_{i-1} of length j $(1 \le j \le |n_{i-1}|)$. So this state is a prefix of v, not ending in a "b".

Fact 4.2. Let S be a semigroup whose idempotents form a band. If e, f are idempotents of S with $e \equiv_R f$ then $e \sim f$.

This follows immediately from the definition of $\sim$.

Actions

To every element $s \in S$ we associate a partial function $Q_v \to Q_v$. The image of $x \in Q_v$ under this map will be denoted $x.s$.

If $v = (n_0,b_1,n_1,\ldots,b_k,n_k)$ then $x \in Q_v$ (a prefix of v) is either of the form $x = (n_0,\ldots,b(r))$ (for $1 \le i \le k$, $r \equiv_R r_i$) or of the form $x = (n_0,\ldots,(a_1{}^i,\ldots,a_j{}^i))$ (for $0 \le i \le k$) where $(a_1{}^i,\ldots,a_j{}^i)$ is a prefix of n_1 of length j (with $1 \le j \le |n_i|$). Let us now define the action of s on x, by cases.

If $x = (n_0,\ldots,b(r))$ with $r \equiv_R r_i$, then

$$x.s = \begin{cases} (n_0,\ldots,b_i,a_1{}^i) & \text{if } b(r) = b_i \text{ and } s = a_1{}^i \text{ (= left-most element of } n_i).\\ (n_0,\ldots,b(r.s)) & \text{if } s \ne a_1^i \text{ and if for all } p \in b(r) \text{ we have } ps \equiv_R p.\\ \text{undefined otherwise.} \end{cases}$$

If $x = (n_0,\ldots,(a_1{}^i,\ldots,a_j{}^i))$ then

$$x.s = \begin{cases} (n_0,\ldots,(a_1{}^i,\ldots,a_j{}^i,a_{j+1}{}^i)) & \text{if } s = a_{j+1}{}^i \text{ and } j+1 < |n_i|.\\ (n_0,\ldots,n_i,b(e_{i+1})) & \text{if } s = a_{j+1}{}^i \text{ and } j+1 = |n_i|, \text{ and } e_{i+1} \text{ is any idempotent with } e_{i+1} \equiv_R r_{i+1}\\ & \text{(Remark. } b(e_{i+1}) \text{ is unique by fact 4.2.)}\\ \text{Undefined otherwise.} \end{cases}$$

If $(s_1,\ldots,s_n) \in S^+$ then we define the action of this word on a state $x \in Q_v$ by $(..(x\ s_1).s_2).\ \ldots).s_k$.

The following properties of the congruence $\sim$ will be used. (We still assume that S is a semigroup whose idempotents form a band.)

Fact 4.3. Let r_1, r_2 be regular elements of S with $r_1 \sim r_2$, and let s be any element of S. Then we have: If $r_1 s \equiv_R r_1$ then $r_2 s \equiv_R r_2$.

A consequence of this fact is that in the definition of the action $x \cdot s$ (in the case where $x = (n_0,\ldots,b(r))$, the condition "for all $p \in b(r)$:

ps $\equiv_R$p" is equivalent to $r \cdot s \equiv_R r$.

Fact 4.4. Let r_1 and r_2 be regular elements of S with $r_1 \equiv_D r_2$, and let s is any element of S, such that $r_1 s \equiv_D r_2 s \equiv_D r_1 \equiv_D r_2$. Then we have:

If $r_1 s \sim r_2 s$, then $r_1 \sim r_2$.

This fact will be used in the proof that the action of s (ε S) on Q_v is a partial injective function.

Proof of facts 4.3 and 4.4: See [B.M.R.].

5) **Proof that the action of s on Q_v is a partial injective function**

Let $v \in R_b(S)$ and $s \in S$ be given and let $x_1, x_2 \in Q_v$ be such that the action of s on x_1 and x_2 is defined and $x_1 \cdot s = x_2 \cdot s$. We must show that then $x_1 = x_2$.

Case 1. $x_1 \cdot s$ is of the form $(n_o, \ldots, (a^i_1, \ldots, a^i_{j+1}))$ (with $1 \leq i \leq k$, $1 \leq j+1 < |n_i|$).
Then, by the definition of the action of s, we must have $s = a^i_{j+1}$ and $x_1 = (n_o, \ldots, (a^i_1, \ldots, a^i_j))$. Similarly, since $x_1 s = x_2 s$ we have $x_2 = (n_o, \ldots, (a^i_1, \ldots, a^i_j))$, hence $x_1 = x_2$.

Case 2. $x_1 \cdot s$ is of the form $(n_o, \ldots, b(r))$ with $r \equiv_R r_i$.

Case 2a. $x_1 = (n_o, \ldots, a^{i-1}_{|n_{i-1}|-1})$ and $s = a^{i-1}_{|n_{i-1}|}$ (then $b(r) = b(e_i)$). Then, since $x_2 s = x_1 s$, we have $x_1 \cdot s = x_2 \cdot s = (n_o, \ldots, s, b(e_i)) = (x_1, s, b(e_i))$. If $x_2 \cdot s = (x_2, s, b(e_i))$ then we have $x_1 = x_2$.

But a priori it also seems possible that x_2 could be of the form $x_2 = (x_1, s, b(r))$ with $rs \sim e_i$. Then we would have $x_1 \neq x_2$ but $x_2 \cdot s = (x_1, s, b(e_i)) = x_1 \cdot s$ (and so s would not act injectively). Let us prove that this cannot arise. Indeed, if $x_1 \cdot s = (x_1, s, b(e_i))$ then the product se_i must be non-regular (because, by definition of the action on Q_v, $(n_o, r_1, \ldots, s, e_i)$ must be in reduced form with respect to R(S); if se_i were regular then $(\ldots, s, e_i)$ could be further reduced to $(\ldots, se_i)$). It follows that $se_i <_L e_i$. Now if we had $rs \sim e_i$ then (multiplying be e_i on the right): $rse_i \sim e_i e_i = e_i$. But this would contradict the fact that $se_i <_L e_i$.

Case 2b. $x_1 = (n_o, \ldots, b(r)) = (y_1, b(r))$ with $r \equiv_R r_i$.

Then, by the reasoning of case 2a, x_2s cannot be of the form $x_2s = (x_2, s, b(e_i))$.

So in this case we have $x_2 = (y_2, b(r'))$ with $r' \equiv_R r_i$. Since s is defined on x_1 and x_2 and $x_1s = x_2s$ we have $y_1 = y_2$ and $b(rs) = b(r's)$. By fact 4.4 this implies $b(r) = b(r')$. Thus $x_1 = x_2$.

6) Proof that non-regular elements of S are not related to group identities

In this section we assume that S is a finite semigroup whose idempotents form a band.

So far we have associated to every element $s \in S$ a set of partial injective functions, each on a set Q_v (where v ranges over the finite set $R_b(S)$).

These partial functions can be extended to permutations of Q_v. Then, using the method of section 3 we obtain a relational morphism of S into the direct product of the symmetric groups $G(Q_v)$ (v ranging over $R_b(S)$).

To show that S_{II} is regular we still must prove that if s is a non-regular element of S then for <u>every</u> factorization $(s_1, \ldots, s_n)$ of s $(= s_1 \ldots s_n)$ there <u>exists</u> $v \in R_b(S)$ such that the action of $(s_1,\ldots,s_n)$ on Q_v is not a subfunction of the identity map of Q_v (i.e. cannot be extended to the identity of $G(Q_v)$). This follows from:

<u>Fact.</u> Let v be the reduced representative of $(s_1, \ldots, s_n)$ in $R_b(S)$ (i.e. first reduce with respect to R(S) then replace the regular coordinates r_i by their blocks $b(r_i)$). Then the action of $(s_1, \ldots, s_n)$ on the start state of Q_v is defined and leads to the state v.

If $s_1, \ldots, s_n$ is non-regular in S then v is different from the start state of Q_v, hence $(s_1, \ldots, s_n)$ does not act as a subfunction of the identity map of Q_v.

<u>Proof.</u> In order to show how $(s_1, \ldots, s_n)$ acts on Q_v we have to compare $(s_1, \ldots, s_n)$ to its reduced representative v in $R_b(S)$. In a semigroup whose regular elements form a subsemigroup (and hence reduced representatives are unique), a sequence $(s_1, \ldots, s_n)$ is reduced by finding maximally long subsegments whose product ($\in S$) is regular. Let $(s_{x_i}, \ldots, s_{y_i})$ be the i-th maximal regular subsegment. Then

$v = (n_0, b_1, n_1, \ldots, b_k, n_k)$ where $n_i = (s_{y_i+1}, \ldots, s_{x_{i+1}-1})$ and

$b_i = b(s_{x_i} \ldots s_{y_i})$.

Then the action of $(s_1, \ldots, s_n)$ on Q_v, starting at the start state is as follows:

First n_o acts, leading to the state $(n_o, b(e_1))$, where e_1 is an idempotent chosen in the R-class of $r_1 = s_{x_1} \ldots s_{y_1}$; $b(e_1)$ is uniquely determined by r_1. Next $(s_{x_1}, \ldots, s_{y_1})$ acts, which consists in successively multiplying e_1 by $s_{x_i}, s_{x_i+1}, \ldots, s_{y_i}$. Since $e_1 \equiv_R e_1 s_{x_1} \equiv_R \cdots \equiv_R e_1 r_1 = r_1$ the action is always defined and the state reached now is $(n_o, b(r_1)) = (n_o, b_1)$. The action of $n_1 = (s_{y_1+1}, \ldots, s_{x_2-1})$ next leads to the state $(n_o, b_1, n_1, b_2(e_2))$. The action continues that way. The state finally reached is $(n_o, b_1, n_1, \ldots, b_k, n_k) = v$.

7) **About the decidability of S_{II}**

In this section we assume that S is a finite semigroup whose regular elements form a subsemigroup (thus the elements of R(S) have unique representatives).

The techniques of sections 4, 5, 6 that deal with non-regular elements lead to a decision procedure which tells whether S_{II} is regular or not.

Definition. The set of strong type II elements of S is the smallest subset of S which:

(1) contains the idempotents of S,

(2) is closed under product (i.e. is a subsemigroup of S),

(3) is closed under weak conjugation (see section 1).

The set of strong type II elements will be denoted $S_{II,strong}$.

Clearly, it is decidable whether an element of S belongs to $S_{II,strong}$ (while at this point it is not known whether there exists an algorithm which, in input S and $s \in S$, decides whether $s \in S_{II}$ or $s \notin S_{II}$).

Theorem. Let S be a finite semigroup whose regular elements form a subsemigroup. Then, S_{II} is regular iff $S_{II,strong}$ is regular.

In addition, "$S_{II,strong}$ is regular" implies $S_{II} = S_{II,strong}$ and it implies that the regular elements of S form a subsemigroup.

Proof. See [B.M.R.].

Acknowledgement: The main incentive for this work was Chris Ash's paper [A]. Tom Hall gave an enlightning presentation of Ash's approach at the Chico Semigroup Conference. We also would like to thank Simon Goberstein and Peter Higgins for organizing this conference.

References

[A] C.J. Ash: Finite semigroups with commuting idempotents, J. Austral. Math. Soc., to appear, (preprint 1985).

[Bi] J.C. Birget: Iteration of expansions, unambiguous semigroups, J. Pure and Appl. Algebra **34** (1984) 1-55.

[B.M.R.] J.C. Birget, S. Margolis, J. Rhodes: Finite semigroups whose idempotents form a semilattice or a band (to be published).

[Br] T.C. Brown: An interesting combinatorial method in the theory of locally finite semigroups, Pacific J. of Math. **36** #2 (1971) 285-289.

[B.R.] J.C. Birget, J. Rhodes: Almost finite expansions of arbitrary semigroups, J. Pure and Appl. Algebra **32** (1984) 239-287.

[H.] J.M. Howie: An Introduction to Semigroup Theory, Acad. Press (1976).

[M.P.] S. Margolis, J.E. Pin:
(1) Expansions, free inverse semigroups, and Schutzenberger product.
(2) Inverse semigroups and extensions of groups by lattices.
(3) Inverse semigroups and varieties of finite semigroups (All three accepted for publication, J. of Algebra.)

[Mc] D.B. McAlister: Regular semigroups, fundamental semigroups and groups, J. Austral. Math. Soc. (Ser. **A**) **29** (1980) 475-503.

[R.T.] J. Rhodes, B. Tilson: Improved lower bounds for the complexity of finite semigroups, J. Pure and Appl. Algebra **2** (1972) 13-71.

[T.W.] D. Thérien, A. Weiss: Varieties of finite categories, RAIRO Info. théor., Vol. **20** #3 (1986).

INVERSE SEMIGROUPS WITH COUNTABLE UNIVERSAL SEMILATTICES

Karl Byleen
Department of Mathematics, Statistics and Computer Science
Marquette University
Milwaukee, WI 53233

ABSTRACT. A semilattice E is said to be a countable universal semilattice if E is countable and if every countable semilattice can be embedded in E. The free Boolean algebra on a countably infinite number of generators is used to construct a particular countable universal semilattice which is the semilattice of idempotents of a 2-generated bisimple monoid.

Even though there are uncountably many pairwise non-isomorphic countable chains (consider the countable ordinals, for example), every countable chain can be embedded in the chain $(Q,\leq)$ of rationals with the usual order. Therefore the chain $(Q,\leq)$, being itself countable, is said to be a countable universal chain. On the other hand, there does not exist a countable universal group, that is, a countable group G in which every countable group can be embedded. For such a group G would have only a countable number of 2-generated subgroups, while by a result of B. H. Neumann [7] there exist uncountably many pairwise non-isomorphic 2-generated groups.

The chain $(Q,\leq)$ has the additional property that any isomorphism between subchains of cardinality $<|Q|$ can be extended to an automorphism of $(Q,\leq)$; we say that $(Q,\leq)$ is homogeneous. In fact, $(Q,\leq)$ is the unique countable homogeneous universal chain. B. Jónsson [5,6] has formulated these notions in the context of relational structures, and has given sufficient conditions for a class of relational structures to possess a homogeneous universal structure of given cardinal. One of these conditions is the (weak) amalgamation property. We refer the reader to Chapter 10 of Bell and Slomson [1] for an exposition of Jónsson's theory.

Imaoka [4] has shown that the varieties of bands which have the (weak) amalgamation property are precisely those contained in the variety of normal bands. It therefore follows from Jónsson's results that any non-trivial variety of normal bands has a unique countable homogeneous universal element. In particular there exists a countable homogeneous universal semilattice. (The countable homogeneous universal left [right] zero band L [R] is just the countably infinite

S. M. Goberstein and P. M. Higgins (eds.), Semigroups and Their Applications, 37–42.

left [right] zero band, and $L \times R$ is the countable homogeneous universal rectangular band).

Our interest in countable universal semilattices is related to embedding theorems for inverse semigroups. N. R. Reilly [9] has shown that any inverse semigroup (and thus, in particular, any semilattice) can be embedded in some bisimple inverse semigroup. Chris Ash has shown that any countable inverse semigroup (and thus any countable semilattice) can be embedded in some 2-generated inverse semigroup. Ash's result appears in the survey article by T. E. Hall [3] in which these embedding theorems and many others are discussed in the context of amalgamation properties for inverse semigroups.

It is unknown whether the results of Reilly and Ash can be combined, that is, whether any countable inverse semigroup can be embedded in some 2-generated bisimple inverse semigroup. (It is known that any countable semigroup can be embedded in a 2-generated bisimple monoid [2]). We will show, however, that any countable semilattice can be embedded in some 2-generated bisimple inverse monoid. In fact we prove a stronger result. We construct a countable universal semilattice which serves as the semilattice of idempotents of a 2-generated bisimple inverse monoid M; thus any countable semilattice can be embedded in M.

Our construction depends upon the duality between Boolean algebras and their Stone spaces, and on certain well-known topological properties of the Cantor set. We recall the relevant results (and refer the reader to [10] and [11] for example, for details) beginning with the Stone representation theorem for Boolean algebras: any Boolean algebra A is isomorphic to the Boolean algebra of all open-closed subsets of some totally disconnected compact Hausdorff space $S(A)$. If A is countable, then the open-closed subsets of $S(A)$ form a countable base and $S(A)$ is metrizable. Thus any countable Boolean algebra A is isomorphic to the Boolean algebra of all open-closed subsets of some totally disconnected compact metric space. In particular, the free Boolean algebra FB_ω on a countably infinite number of generators is isomorphic to the Boolean algebra of all open-closed subsets of the Cantor set $\mathcal{C}$. But every compact metric space is a continuous image of $\mathcal{C}$, and thus every countable Boolean algebra is embeddable in FB_ω.

The Cantor set $\mathcal{C}$ is, up to homeomorphism, the only totally disconnected perfect compact metric space. In particular, any non-empty open-closed subset G of $\mathcal{C}$ is homeomorphic to $\mathcal{C}$. A homeomorphism from $\mathcal{C}$ to G induces a Boolean isomorphism from FB_ω onto the Boolean algebra of all open-closed subsets of G (i.e., onto the principal ideal of FB_ω generated by G). Furthermore, if G and H are any open-closed subsets of $\mathcal{C}$ which are proper and non-empty, then G and H, as well as $\mathcal{C} - G$ and $\mathcal{C} - H$, are homeomorphic, and so there exists a homeomorphism of $\mathcal{C}$ which maps G onto H. This homeomorphism induces a Boolean automorphism of FB_ω which takes G to H.

The 2-generated bisimple inverse semigroup we seek will be obtained as a full inverse subsemigroup of the Munn semigroup T_E for an appropriate semilattice E. (For standard results and terminology concerning inverse semigroups, including an exposition of the theory of fundamental inverse semigroups due to W. D. Munn, we refer the reader to Petrich [8]). We single out as our first lemma a construction of a certain principal ideal isomorphism of a semilattice. Let C be a chain and let $\{S_\alpha : \alpha \in C\}$ be a collection of pairwise disjoint semigroups. Let $S = \bigcup_{\alpha \in C} S_\alpha$ and if $s \in S_\alpha$, $t \in S_\beta$ define

$$s \cdot t = \begin{cases} s & \text{if } \alpha < \beta \\ st & \text{if } \alpha = \beta \\ t & \text{if } \beta < \alpha \end{cases}$$

Then S is a semigroup, said to be *the chain* C *of the semigroups* $\{S_\alpha : \alpha \in C\}$. We note that if each S_α is a semilattice, then so is S, and if each S_α is isomorphic to the semilattice L, then S is isomorphic to the ordinal product $C \circ L$ of the semilattice C with the semilattice L.

LEMMA 1. Let E denote the chain C of the semilattices $\{E_\alpha : \alpha \in C\}$. If $\phi_\alpha : E_\alpha e \to E_\alpha f$ is a principal ideal isomorphism of E_α and if $\phi_\beta : E_\beta \to E_\beta$ is an automorphism whenever $\beta < \alpha$, then the mapping $\theta : Ee \to Ef$ defined by

$$x\theta = \begin{cases} x\phi_\alpha & \text{if } x \le e \text{ in } E_\alpha \\ x\phi_\beta & \text{if } x \in E_\beta \text{ for some } \beta < \alpha \end{cases}$$

is a principal ideal isomorphism of E.

PROOF: Since $Ee = (\bigcup_{\beta<\alpha} E_\beta) \cup E_\alpha e$ and $Ef = (\bigcup_{\beta<\alpha} E_\beta) \cup E_\alpha f$ the mapping $\theta : Ee \to Ef$ is a bijection. Let $x,y \in Ee$, say $x \in E_\gamma$ and $y \in E_\delta$. If $\gamma = \delta$, then $(xy)\theta = (xy)\phi_\gamma = x\phi_\gamma \cdot y\phi_\gamma = x\theta \cdot y\theta$ since ϕ_γ is an isomorphism. If $\gamma \neq \delta$, then either $\gamma < \delta$ or $\delta < \gamma$. In the first case $(xy)\theta = x\theta = x\phi_\gamma = x\phi_\gamma \cdot y\phi_\delta = x\theta \cdot y\theta$ and the second case is similar. Hence θ is a principal ideal isomorphism.

We consider FB_ω to be represented as the set of all open-closed subsets of the Cantor set C. Let F denote the semilattice obtained from FB_ω by deleting its maximum element C and considering only the meet operation (intersection). Let $C_\omega = \{0,1,2,...\}$ denote the ω-chain of non-negative integers with the reverse of the usual order. Let $F_0 = \{0\} \times F^1$ and let $F_n = \{n\} \times F$ for $n = 1,2,3,\ldots$. Finally, let E denote the chain C_ω of semilattices $\{F_n : n \in C_\omega\}$.

LEMMA 2. E is a countable universal semilattice.

PROOF: It is clear that E is countable. To show that E is universal, let L be a countable semilattice. The mapping $\psi : L \to 2^L$ defined by $x \to \{z \in L : z \leq x\}$ embeds L in the (uncountable) semilattice 2^L of all subsets of L under intersection. Now 2^L may be considered a Boolean algebra in the usual way. The sub-Boolean algebra $\langle\psi(L)\rangle$ of 2^L which is generated by $\psi(L)$ is countable. Thus, by our earlier remarks, $\langle\psi(L)\rangle$ is embeddable in FB_ω, and so the semilattice L is embeddable in F^1, and thus in E.

Let $\alpha : E \to E(0,\emptyset)$ be defined by

$$(n,x) \to \begin{cases} (0,\emptyset) & \text{if } (n,x) = (0,1) \\ (n+1,x) & \text{otherwise.} \end{cases}$$

Again we make use of our earlier remarks concerning Boolean algebras and the Cantor set to define a mapping $\beta : E \to E$. Let $R_1, R_2, R_3, \ldots$ be an enumeration of the non-empty elements of F. For each $i = 1,2,3,\ldots$ there exists a Boolean automorphism of FB_ω taking R_1 to R_i, and thus a semilattice automorphism $\beta_i : F \to F$ such that $R_1\beta_i = R_i$. There also exists a Boolean isomorphism of FB_ω onto the principal ideal $\langle R_1\rangle$ of FB_ω generated by R_1, and thus a semilattice isomorphism $\beta_0 : F^1 \to \langle R_1\rangle$. Define $\beta : E \to E(0,R_1)$ by $(n,x) \to (n,x\beta_n)$.

LEMMA 3. The mappings $\alpha : E \to E(0,\phi)$ and $\beta' : E \to E(0,R_1)$ are principal ideal isomorphisms of E.

PROOF: Since $E = \{(0,1)\} \cup \{(n,x) : n \geq 0, x \in F\}$ and $E(0,\phi) = \{(0,\phi)\} \cup \{(n,x) : n \geq 1, x \in F\}$ the mapping α is a bijection. Let (n,x), $(m,y) \in E$. To show that α is a homomorphism we first assume that neither (n,x) nor (m,y) equals $(0,1)$ and suppose $n < m$ (in C_ω). Then $[(n,x)(m,y)]\alpha = (n,x)\alpha = (n+1,x) = (n+1,x)(m+1,y) = (n,x)\alpha\cdot(m,y)\alpha$. The case $m < n$ is similar. If $n = m$ then $[(n,x)(m,y)]\alpha = (n,xy)\alpha = (n+1,xy) = (n+1,x)(n+1,y) = (n,x)\alpha(m,y)\alpha$. We easily draw the same conclusion if one or both of $(n,x),(m,y)$ equals $(0,1)$, and so α is a homomorphism and therefore a principal ideal isomorphism of E. The mapping β is precisely the mapping θ of Lemma 1 which is constructed from the principal ideal isomorphism $\phi_0 : F_0 \to F_0$ defined by $(0,x) \to (0,x\beta_0)$ and the semilattice automorphisms $\phi_n : F_n \to F_n$ defined by $(n,x) \to (n,x\beta_n)$ for $n = 1,2,3,\ldots$.

Therefore, by Lemma 1, β is also a principal ideal isomorphism of E.

LEMMA 4. Let $(n,x) \in E$. Then there exists a principal ideal isomorphism γ belonging to the inverse subsemigroup $\langle\alpha,\beta\rangle$ of T_E generated by α and β which maps $E(0,1)$ onto $E(n,x)$.

PROOF: Let $(n,x) \in E$. If $x = 1$, then $(n,x) = (0,1)$ and we may let $\gamma = \alpha\alpha^{-1}$. If $x = \emptyset$, let $\gamma = \alpha^{n+1}$. Then $(0,1)\alpha^{n+1} = (n,\emptyset)$, as required. Finally suppose $x \neq 1$ and $x \neq \emptyset$. Then $x = R_i$ for some positive integer i. Let $\gamma = \beta\alpha^{i}\beta\alpha^{n-i}$. Then $(0,1)\gamma =$ $(0,1)\beta\alpha^{i}\beta\alpha^{n-i} = (0,R_1)\alpha^{i}\beta\alpha^{n-i} = (i,R_1)\beta\alpha^{n-i} = (i,R_i)\alpha^{n-i} = (n,R_i)$, so α again maps $E(0,1)$ onto $E(n,x)$.

THEOREM 5. There exists a 2-generated bisimple inverse monoid in which every countable semilattice can be embedded.

PROOF: By Lemma 4 the inverse subsemigroup $\langle\alpha,\beta\rangle$ of T_E is transitive, so $\langle\alpha,\beta\rangle$ is bisimple and full (i.e., contains E). So by Lemma 2 every countable semilattice can be embedded in $\langle\alpha,\beta\rangle$.

At the beginning of the paper we noted that there is no countable universal group. Similarly there does not exist a countable universal inverse semigroup, so Theorem 5 cannot be strengthened by replacing "semilattice" by "inverse semigroup". Furthermore, "2-generated" cannot be replaced by "monogenic" since any monogenic bisimple inverse semigroup is isomorphic to either the bicyclic semigroup or to a cyclic group, as can be deduced from the classification of monogenic inverse semigroups [8]. The only semilattices which can therefore be embedded in monogenic bisimple inverse semigroups are the finite chains and the ω-chain.

In conclusion we pose some problems and questions suggested by our results.

1. Find concrete representations for the countable homogeneous universal elements of the varieties of normal bands which contain semilattices.

2. Can every countable inverse semigroup be embedded in some 2-generated [finitely generated] bisimple inverse semigroup?

3. Can every countable [inverse] semigroup be embedded in some finitely generated congruence-free [inverse] semigroup?

REFERENCES

1. J. L. Bell and A. B. Slomson, <u>Models and ultraproducts: an introduction</u>, North Holland, Amsterdam, 1971.
2. K. Byleen, 'Embedding any countable semigroup in a 2-generated bisimple monoid', Glasgow Math. J. <u>25</u> (1984), 153-161.
3. T. E. Hall, 'Inverse and regular semigroups and amalgamation: a brief survey', Proceedings of a Symposium on Regular Semi-Groups, Northern Illinois University, 1979.

4. T. Imaoka, 'Free products with amalgamation of bands', Mem. Fac. Lit. & Sci., Shimane Univ., Nat. Sci. 10 (1976), 7-17.
5. B. Jónsson, 'Universal relational systems', Math. Scand. 4 (1956), 193-208.
6. B. Jónsson, 'Homogeneous universal relational systems', Math. Scand. 8 (1960), 137-142.
7. B. H. Neumann, 'Some remarks on infinite groups', J. London Math. Soc. 12 (1937), 120-127.
8. M. Petrich, Inverse Semigroups, Wiley, New York, 1984.
9. N. R. Reilly, 'Embedding inverse semigroups in bisimple inverse semigroups', Quart. J. Math., Oxford (2) 16 (1965), 183-187.
10. R. Sikorski, Boolean algebras, Springer-Verlag, Berlin, 1964.
11. S. Willard, General topology, Addison-Wesley, Reading, 1970.

THE IDENTITY ELEMENT IN INVERSE SEMIGROUP ALGEBRAS

Michael P. Drazin
Department of Mathematics
Purdue University
West Lafayette, Indiana 47907
U.S.A.

ABSTRACT. W. D. Munn and R. Penrose (1957) obtained an explicit formula for the identity element 1_A of the semigroup algebra A of an arbitrary finite inverse semigroup S. An alternative (inductive) characterization of 1_A is presented, giving new information about how the form of 1_A depends on the Vagner-Preston order in S.

1. INTRODUCTION

Let $S = \{a_1, \ldots, a_n\}$ be any given finite semigroup, let

$$A = Q[S] = \{b = \beta_1 a_1 + \ldots + \beta_n a_n : \beta_i \in Q\}$$

be the semigroup algebra of S over the field Q of rational numbers, and let $N = N(A)$ denote the radical of A. Then the quotient algebra A/N is semisimple, and, by standard Wedderburn-Artin theory, consequently has a unique two-sided identity element, say

$$1_{A/N} = (\gamma_1 a_1 + \ldots + \gamma_n a_n) + N.$$

In general of course $N \neq 0$, so that $\gamma_1, \ldots, \gamma_n$ are not uniquely determined. For simplicity, suppose for the present that $N = 0$, in which case $\gamma_1, \ldots, \gamma_n$ become rational invariants of S, i.e. we have a canonical map $k: S \to Q$, which we denote as $a_i \mapsto k_{a_i} = \gamma_i$.

In particular, certainly $N = 0$ when S is inverse (see e.g. [2, Theorem 4.4, p.9]), and in this case R. Penrose (see [2, Section 4.9, p.12]) found that the identity element of A is given by the formula

$$1_A = s_1 - s_2 + \ldots + (-1)^{m-1} s_m,$$

where s_r $(r = 1, 2, \ldots, m)$ denotes the sum of all products of r distinct idempotents maximal with respect to the usual partial ordering of the semilattice $E = E(S)$ of idempotents in S, and where m denotes the number of

S. M. Goberstein and P. M. Higgins (eds.), Semigroups and Their Applications, 43–46.

these maximal idempotents. As well as giving 1_A, this formula tells us that (for inverse S) the map k vanishes outside E and takes only integer values, so we may as well regard k as a map $k: S \to Z$.

My own interest in this map arose not so much in k for its own sake, but rather because the connection between k and the semilattice E(S) for inverse S seemed to suggest that, if one could first somehow find k in some other sufficiently explicit way for non-inverse S, one might then be able to use k to infer the existence of some new natural ordered structure on S (or on some subset of S).

With this objective in view, it is desirable first to have the action of k, in the inverse case, expressed as directly as possible in terms of the order on E, and it turns out that there is indeed another way of describing k.

2. METHODS AND RESULTS

We start by regarding k as a labelling of the Hasse diagram of the semilattice E as in Figure 1 (which shows two more or less randomly chosen illustrative examples, with vertex labels computed from the Munn-Penrose

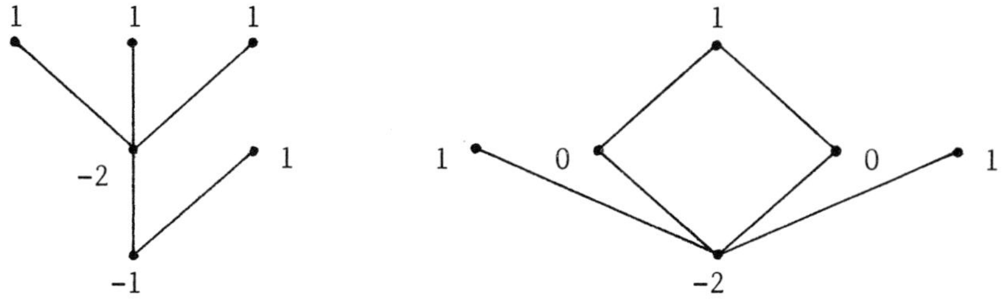

Figure 1. The k-labelling of two semilattices of order 6

formula). Quite generally, for any finite inverse S, the Munn-Penrose formula tells us that

$$k_e = 1 \quad \text{whenever} \quad e \quad \text{is maximal in} \quad E,$$

and, after examining a few concrete examples such as those in Figure 1, it is easy to conjecture that the labels for non-maximal idempotents e can then be obtained inductively, in terms of the length of the longest chain in E between e and a maximal idempotent, by the formula *

$$k_e = 1 - \sum\{k_g : g \in E \text{ and } g > e\}.$$

* This inductive definition of k_e is close to that of the Möbius function, which differs only in omitting the first term 1 on the right-hand side (see e.g. [3, p. 344], and cf. also [1, pp. 328-332]).

This is indeed valid as an alternative description of the map k. Presumably it can be deduced from the Munn-Penrose formula, but my own approach was somewhat different. For any $e, f \in E$, define a subset

$$E_{e,f} = \{g: g \in E \text{ and } e = fg\}$$

of E (non-empty iff $e \leq f$), and an integer

$$\sigma_{e,f} = \sum\{k_g: g \in E_{e,f}\};$$

if $E_{e,f}$ is empty, then we interpret this as meaning $\sigma_{e,f} = 0$.

PROPOSITION 1. $\sigma_{e,f} = \begin{cases} 1 & \text{if } e = f \\ 0 & \text{if } e \neq f. \end{cases}$

The proof of this is (of course) accomplished by an inductive argument, the details of which will appear elsewhere. Assuming Proposition 1, it is now an easy matter to establish the desired property of the inductively defined map k:

PROPOSITION 2. $u = \sum_{g \in E} k_g g$ _is the two-sided identity element of_ Q[E].

Proof. For any $f \in E$, we have

$$\begin{aligned} fu &= f.(\sum_g k_g g) = \sum_g k_g fg \\ &= \sum_{e \in E} \left(\sum \left\{ k_g: g \in E_{e,f} \right\} \right) e = \sum_{e \in E} \sigma_{e,f} e = f. \end{aligned}$$

PROPOSITION 3. _If_ S _is any finite inverse semigroup having_ E _as its semilattice of idempotents, then_ $u = \sum_{g \in E} k_g g$ _is also the two-sided identity element of_ Q[S].

Proof (from [2, Section 4.9, p.12]). For any $a \in S$, we have

$$ua = u.aa^{-1}a = (u(aa^{-1}))a = (aa^{-1})a = a,$$

and similarly $au = a$.

3. REMARKS

For arbitrary finite S there is no longer any obvious semilattice (or indeed any obvious ordered set) available, but one can still always compute the two-sided identity element of A/N directly. And, surprisingly, computer analysis of semigroups of small finite order n revealed that for arbitrary semigroups with $n \leq 6$, one can always choose all the coefficients k_g in

$$1_{A/N} = \left(\sum_{g \in S} k_g g \right) + N$$

as integers, which suggested the possibility of using these k_g's (at least when $N = 0$) to infer some order structure in S.

However, this seems to be only a mirage. When I communicated the observations above to Munn, he suggested that, for $n = 10$, the 0-simple semigroup $M^0\left(\{1\};\ 3,\ 3;\ \begin{pmatrix} 0 & 1 & 1 \\ 1 & 0 & 1 \\ 1 & 1 & 0 \end{pmatrix}\right)$ should provide non-integer k_g's; and in fact one easily finds

$$k_0 = -1/2,$$

$$k_{(i,\lambda)} = \begin{cases} -1/2 & \text{when } i = \lambda \\ +1/2 & \text{when } i \neq \lambda. \end{cases}$$

Thus our hoped-for avenue to finding a natural order structure valid for arbitrary finite semigroups seems to be only a dead end.

REFERENCES

1. G. LALLEMENT, Semigroups and Combinatorial Applications, Wiley, New York, 1979.

2. W. D. MUNN, 'Matrix representations of semigroups', Proc. Cambridge Phil. Soc. 53 (1957), 5-12.

3. G.-C. ROTA, 'On the foundations of combinatorial theory I. Theory of Möbius functions', Zeit. Wahrscheinlichkeit 2 (1964), 340-368.

FREE BANDS AND FREE *-BANDS

J. A. Gerhard
Department of Mathematics and Astronomy
University of Manitoba
Winnipeg, Canada
R3T 2N2

This talk is based on a paper appeared in the July 1986 issue of the Glasgow Mathematical Journal. It is joint work with Mario Petrich.

1. FREE BANDS

The solution of the word problem for free bands dates back to 1954 (Green and Rees [2]). The usual solution is inductive on the number of variables. I am going to describe a variant of this solution. It gives a canonical representation for words. Another canonical representation by words of shortest length was given by Siekmann and Szabo [3].

Let X be a set and let $F(X)$ be the free semigroup on X. Let $FB(X)$ be the free band on X. Then $FB(X) \cong F(X)/\beta$ where β is the fully invariant congruence on $F(X)$ generated by $x = x^2$.

To describe β and thereby give a solution to the word problem for $FB(X)$ we need some notation. Let $u,v,p,q,w \in F(X)$ and $x,y \in X$. The <u>content</u> of w, $c(w)$ is the set of variables occurring in w. Let $w = uxp$ where $c(w) = c(ux)$ and $c(w) \neq c(u)$. Then

$$s(w) = u$$
$$\sigma(w) = x.$$

(Note that if $c(w) = \{x\}$ then $s(w)$ is the empty word.) Dually let $w = qyv$ where $c(w) = c(yv)$ and $c(w) \neq c(v)$. Then

$$\varepsilon(w) = y$$
$$e(w) = v.$$

The usual solution for the word problem for $FB(X)$ is as follows:

$$u\beta v \iff \begin{cases} c(u) = c(v) \\ s(u)\beta s(v) \text{ and } e(u)\beta e(v). \end{cases}$$

S. M. Goberstein and P. M. Higgins (eds.), Semigroups and Their Applications, 47–50.

This is inductive on $|c(u)|$ since $|c(s(u))| + 1 = |c(u)|$. We give a solution to the word problem by defining $b : F(X) \to F(X)$ so that $FB(X) \cong b(F(X))$ and with multiplication $u \circ v = b(uv)$ where uv is the product in $F(X)$. Define

$$b(w) = b(s(w))\sigma(w)\varepsilon(w)b(e(w)).$$

Of course b is defined by induction on the number of variables in w. The induction starts with $b(x^n) = xx$ for any $n \geq 1$. As an example, if

$$w = (xy\ xy^3\ xz)^2 y$$

then

$$b(w) = b(xy\ xy^3\ x)\ zx\ b(zy)$$

$$= b(x)\ yy\ b(x)\ zx\ b(z)\ yz\ b(y)$$

$$= xx\ yy\ xx\ zx\ zz\ yz\ yy$$

The length of $b(w)$ is $2(2^{|c(w)|}-1)$.
It can be proved that

$$FB(X) \cong b(F(X)) = \{w | w = b(w)\}$$

and the canonical words are the words of the form $w = b(w)$.

Our canonical form for a word reflects the usual inductive solution of the word problem for bands. Another canonical form was given by Siekmann and Szabo [3] where they show that each β-class contains a unique shortest word. For the above example this word is $xyxzy$. This shortest word can be obtained from $b(w)$ by replacements of p^2 by p. In general, replacements of upv by uv in case $c(p) \subseteq c(u) = c(v)$ are also necessary.

2. FREE *-BANDS

We give a solution of the word problem for free *-bands which is analogous to our solution for free bands. The concept of the "content" for free *-bands was known to Adair [1]. Also Yamada [4] has discussed the size of finitely generated free *-bands.

Free *-bands are studied in the context of involutorial semigroups. An <u>involutorial semigroup</u> is a semigroup with a unary operation $*$ which satisfies

$$x^{**} = x$$

$$(xy)^* = y^* x^*.$$

The <u>free involutorial</u> semigroup on X, $F^*(X)$ is defined as follows. Let $I = X \cup X^*$ where X^* is disjoint from X and in one-to-one correspondence with X via $x \to x^*$. Then

$$F^*(X) = (F(I),*)$$

where $F(I)$ is the free semigroup on I and $*$ is defined on $F(I)$ by extending $* : I \to I$ freely to an antiautomorphism of $F(I)$ of degree 2 (an involution). The <u>free *-band</u> on X, $FB^*(X)$, is isomorphic to $F(X)/\beta^*$ where β^* is defined by the equations $x = x^2$ and $x = xx^*x$.

As in the case of the free band, to describe a solution of the word problem for $FB^*(X)$, we define $b^* : F^*(X) \to F^*(X)$ so that $FB^*(X) \cong b^*(F^*(X))$. To define b^* we need the following notation. Let $u,v,p,q,w \in F^*(X)$ and $i,j, \in I$. Let

$$c_X(w) = \{x \in X \mid x \text{ or } x^* \text{ occurs in } w\}.$$

This is a reasonable definition of "content" because of the equation $x = xx^*x$. Let $w = uip$ where $c_X(w) = c_X(ui)$ and $c_X(w) \neq c_X(u)$. Then

$$s_X(w) = u$$

$$\sigma_X(w) = i.$$

We can do without the analogues of ε and e because of $*$. Define

$$b^*(w) = b^*(s_X(w))\ \sigma_X(w)[b^*(s_X(w^*))\ \sigma_X(w^*)]^*.$$

Then

$$FB^*(X) \cong b^*(F^*(X)) = \{w \mid w = b^*(w)\}$$

where

$$u \circ v = b^*(uv)$$

and

$$(b^*(w))^* = b^*(w^*).$$

3. RELATION BETWEEN b AND b^*.

Let $FIB(X)$ be the free involutorial band on X. (This is just the quotient of $F^*(X)$ by the congruence generated by $x = x^2$.) Now $FIB(X) \cong (FB(I),*)$ where $*$ is defined on $FB(I)$ in the same way as it was defined on $F(I)$. The relation between b and b^* is given by the following commutative diagram.

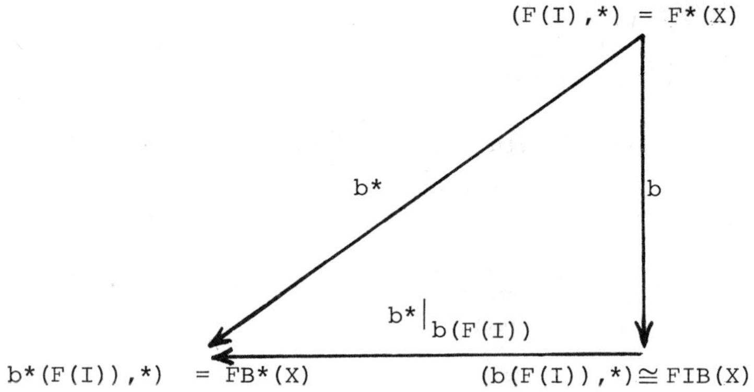

The semigroup homomorphism b also preserves *. The homomorphism $b^*|_{b(F(I))}$ induces the congruence generated by $x = xx^*x$ on FIB(X).

References

[1] Adair, C. L., 'Bands with involution', J. Algebra 75 (1982), 297-314.

[2] Green, J. A. and D. Rees, 'On semi-groups in which $x^r = x$', Proc. Cambridge Phil. Soc. 48 (1952), 35-40.

[3] Siekmann, J. and P. Szabo, 'A noetherian and confluent rewrite system for idempotent semigroups', Semigroup Forum 25 (1982), 83-110.

[4] Yamada, M., 'Finitely generated free *-bands', Semigroup Forum 29 (1984), 13-16.

FINITE INVERSE SEMIGROUPS AND AMALGAMATION

T.E. Hall
Mathematics Department
Monash University
Clayton
Victoria 3168
Australia

ABSTRACT. Given that the class of all inverse semigroups has the (strong) amalgamation property, and that the class of finite inverse semigroups does not, we consider the question of which inverse semigroups are amalgamation bases in the latter class, and we find that they are precisely finite inverse semigroups whose partially ordered sets of $\mathcal{J}$-classes are chains. Some background results are also given.

INTRODUCTION

For our purposes, we can regard an <u>amalgam</u> to be a list $(S,T;U)$ of semigroups such that U is a subsemigroup of S and of T. A more general definition is given in [1], [9] and [10].

We say that the amalgam $(S,T;U)$ is <u>weakly embeddable</u> in a semigroup P if there are monomorphisms $\varphi: S \to P$, $\psi: T \to P$ which agree on U; if further $S\varphi \cap T\psi = U\varphi (= U\psi)$ then we say that $(S,T;U)$ is <u>strongly embeddable</u> in P. If every amalgam of semigroups from a class $\mathcal{C}$ of semigroups is weakly [strongly] embeddable in some semigroup $P \in \mathcal{C}$, then $\mathcal{C}$ is said to have the <u>weak</u> [<u>strong</u>] <u>amalgamation property</u>. O. Schreier's classic result [11] can be stated as "the class of groups has the strong amalgamation property".

In 1957, Kimura (see [1, Section 9.4]) showed that the class of all semigroups has not the weak (and hence not the strong) amalgamation property : the following example is a slight modification of his, in that it contains one fewer element. Let U be the three element null semigroup with elements $0,u,v$ say (thus $xy = 0$ for all $x,y \in U$).

Let $S = U \cup \{a\}$ be the semigroup with U as a subsemigroup and satisfying $a^2 = 0$, $au = ua = v$, $av = va = 0$. Let $T = U \cup \{b\}$ be the semigroup with U as a subsemigroup and such that $b^2 = 0$, $bv = vb = u$, $bu = ub = 0$. From the equations $0 = 0b = (av)b = a(vb) = au = v$ it is easy to see that the amalgam $(S,T;U)$ is not weakly embeddable in any semigroup.

S. M. Goberstein and P. M. Higgins (eds.), Semigroups and Their Applications, 51–56.

The author [3] showed that the class of inverse semigroups has the strong amalgamation property (see [10] and [11] for expositions). The proof in [3] was via two properties of inverse semigroups: the right congruence extension property (for any right congruence δ on any inverse subsemigroup U of any inverse semigroup S, there is a right congruence σ on S such that $\sigma \cap (U \times U) = \delta$) and the representation extension property for one-to-one partial transformations (for any inverse subsemigroup U of any inverse semigroup S and any morphism $\alpha : U \to \mathcal{I}(X)$ (here $\mathcal{I}(X)$ denotes the symmetric inverse semigroup of all one-to-one partial transformations of any set X) there exist a set Y and a morphism $\beta : S \to \mathcal{I}(X \cup Y)$ such that $\beta_u | X = \alpha_u$ for all $u \in U$).

In the survey article [6] on amalgamation, the author gave many applications of the above result that the class of inverse semigroup has the strong amalgamation property. A recent application to "essential conjugate extensions" is given in [11].

A still unpublished joint result of the author and T. Imaoka is that the class of generalized inverse semigroups (othodox semigroups whose bands are normal) also has the strong amalgamation property.

C.J. Ash's example given in [3] shows that the class of all finite inverse semigroups does not have the weak amalgamation property. Let U be the three element semilattice $\{0,e,f\}$ which is not a chain, and with zero element 0.

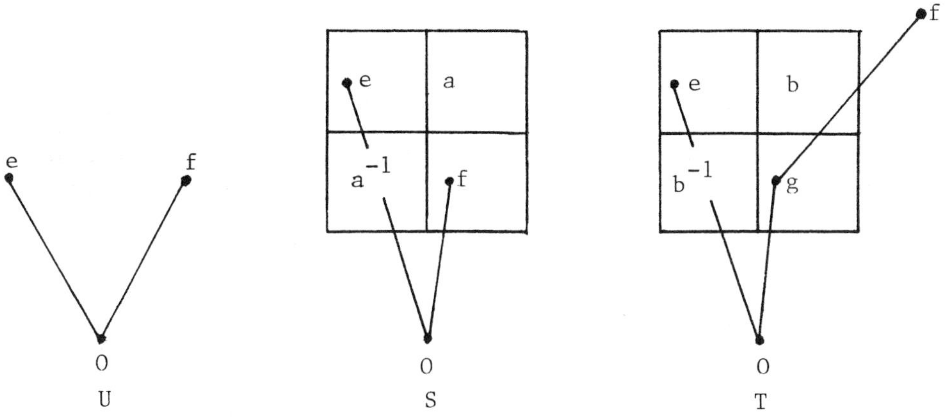

Put $S = T_U$, the Munn semigroup of U: that is, S is the five element combinatorial Brandt semigroup. Put $E = U \cup \{g\}$, a semilattice containing U as a subsemilattice and such that $eg = 0$, $fg = g$. Now put $T = T_E$, the Munn semigroup of E. Then the amalgam $(S,T;U)$ is not weakly embeddable in any finite semigroup: for suppose there exist a semigroup P and monomorphisms $\varphi : S \to P$, $\psi : T \to P$ such that $u\varphi = u\psi$ for all $u \in U$. Then $g \mathcal{D} e$ in T and $e \mathcal{D} f$ in S so $g\psi \ \mathcal{D} \ f\psi$ in P; also $g \leqslant f$ in T so $g\psi \leqslant f\psi$ in P. Thus P contains a copy of the bicylic semigroup [1, Theorem 2.54] and so is infinite, as required.

AMALGAMATION BASES

A weak [strong] amalgamation base for a class $\mathcal{C}$ of semigroups is a semigroup $U \in \mathcal{C}$ such that every amalgam of the form (S,T;U) from $\mathcal{C}$ is weakly [strongly] embeddable in a semigroup $P \in \mathcal{C}$. Grätzer and Lakser [2] define the amalgamation class of a class $\mathcal{C}$ to be the class of amalgamation bases of $\mathcal{C}$.

For $\mathcal{C}$ the class of all semigroups, the amalgamation class contains all inverse semigroups (Howie [9]), all finite cyclic semigroups, and the two element left [right] zero semigroup (the author [5, Theorem 25]): it does not contain the infinite cyclic semigroup, nor the three element null semigroup (Kimura; see [1, Section 9.4]). Also it does not contain the n element left [right] zero semigroup with $n \geqslant 3$ (the author [5, Theorem 25]) from which it is easy to see that the largest class of regular semigroups, each an amalgamation base in the class of all semigroups, which is closed under direct products and regular subsemigroups, is the class of inverse semigroups. (The corresponding question was independently asked by J.M. Howie.)

For the class $\mathcal{C}$ of all semigroups, each weak amalgamation base is also a strong amalgamation base [5, Theorem 12]. (This is not true of the class of distributive lattices, which has the weak but not the strong amalgamation property [2] .)

In the next section we turn to the question of finding the amalgamation class of the class of all finite inverse semigroups. We end this section by noting an amalgamation result in which finiteness can be preserved. Let (S,T;U) be an amalgam of regular semigroups such that U is full in S and in T (i.e. such that E(U) = E(S) = E(T)). Then (S,T;U) is strongly embeddable in a regular semigroup P such that E(U) = E(P), and finiteness can be preserved. Since E(U) = E(S) = E(T) = E(P), the properties of being inverse, or orthodox for example, can also be preserved.

This result played an unexpected and major role in proving the following result in [7] : any regular semigroup S is embeddable in a regular semigroup T such that (i) S is full in T, i.e. E(S) = E(T), and (ii) for all $e,f \in E(S)$, $e \mathcal{D} f$ in T if (and only if) $\langle e \rangle$ is isomorphic to $\langle f \rangle$ (where $\langle e \rangle = \langle E(eSe) \rangle$, the subsemigroup generated by the idempotents of eSe).

This embedding result had seven corollaries in [7], finally giving characterizations, for example, of the core (the idempotent generated subsemigroup) of a regular simple [bisimple] semigroup. One result was that an idempotent generated regular semigroup is the core of a bisimple regular semigroup if and only if it is uniform (i.e. $\langle e \rangle \cong \langle f \rangle$ for all idempotents e,f).

AMALGAMATION BASES FOR FINITE INVERSE SEMIGROUPS

The following result tells us when two elements in a finite inverse semigroup can be made $\mathcal{J}$-related in a containing finite inverse semigroup. (A much stronger result actually appears in [8].)

THEOREM 1 (Hall and Putcha [8]). *Take any elements* a,b *in a finite inverse semigroup* S. *Then there exists a finite inverse semigroup* T *containing* S *such that* $a \mathcal{J} b$ *in* T *if and only if either* $a \mathcal{J} b$ *in* S *or* $J_a \nleq J_b$ *and* $J_b \nleq J_a$ *in* S.

A combination of this theorem and the proof above that Ash's example of an amalgam of finite inverse semigroups is not embeddable in a finite semigroup, gives us a necessary condition for a finite inverse semigroup to be an amalgamation base in the class of finite inverse semigroups. (The condition is also valid for the class of finite [regular] semigroups [8, Theorem 3].)

THEOREM 2 (Hall and Putcha [8]). *If* U *is an amalgamation base in the class of finite inverse semigroups then* $U/\mathcal{J}$ *is totally ordered.*

PROOF. Suppose that U is a finite inverse semigroup such that $U/\mathcal{J}$ is not totally ordered, and take two noncomparable $\mathcal{J}$-classes in U, and choose any idempotents e,f, one from each of those two $\mathcal{J}$-classes. Form an inverse semigroup $U \cup \{g\}$ (where $g \notin U$) containing U as a subsemigroup and such that $g^2 = g$, $gu = fu$, $ug = uf$ for all $u \in U$ (then $U \cup \{g\}$ is an ideal extension of U by $\{g\}^0$ determined by the partial homomorphism $g \mapsto f$ [1, Theorem 4.19]).

By Theorem 1 there exists a finite inverse semigroup S containing U such that $e \mathcal{J} f$ in S. It is easily seen that J_e and $J_g = \{g\}$ are not comparable in $U \cup \{g\}$ so again by Theorem 1 there exists a finite inverse semigroup T containing $U \cup \{g\}$ such that $e \mathcal{J} g$ in T. As before, we show that (S,T;U) is not embeddable in any finite semigroup, for if $\varphi : S \to P$, $\psi : T \to P$ are monomorphisms into a semigroup P such that $\varphi|U = \psi|U$, then $e \mathcal{J} f$ in S, $f < g \mathcal{J} e$ in T give us that $f\psi < g\psi$ and $f\psi \mathcal{J} g\psi$ in P, whence P contains the bicyclic semigroup, and so is infinite. The result follows.

THE CONVERSE RESULT

Recently the author has proved the converse result, namely

THEOREM 3. *If* U *is a finite inverse semigroup such that* $U/\mathcal{J}$ *is totally ordered, then* U *is an amalgamation base in the class of finite inverse semigroups.*

We shall give two key results involved in the proof of this theorem : full proofs will appear elsewhere.

First we recall the definition of orbits, and of isomorphic orbits, for representations of inverse semigroups by one-to-one partial transformations. Let $\rho: U \to \mathcal{I}(X)$ be any morphism of an inverse semigroup $U = U^1$ (without loss of generality) into the symmetric inverse semigroup $\mathcal{I}(X)$ on any set X such that $\rho_1 = 1_X$. An orbit of ρ is a subset 0 of X such that, for some $x \in 0$,

$$0 = \{x\rho_u : u \in U\}.$$

In fact U acts transitively on 0 via ρ, i.e. for every $y \in 0$, $0 = \{y\rho_u : u \in U\}$, and X is a disjoint union of the orbits of ρ.

For any other representation $\rho' : U \to \mathcal{I}(X')$ say, and any orbit $0'$ of ρ', we call a bijection $\varphi: 0 \to 0'$ an isomorphism of orbits if $x\rho_u\varphi = x\varphi\rho'_u$ for all $x \in X$.

For U any inverse semigroup, each $\mathcal{R}$-class R of U, and each subgroup K of the maximal subgroup G of R, yield an orbit as follows. Define

$$0 = R/K = \{Kr : r \in R\},$$

a disjoint union of subsets of R, called the cosets of K in R.

Define $\rho : U \to \mathcal{I}(R/K)$ by

$$(Kr)\rho_u = \begin{cases} Kru & \text{if } ru \in R \\ \text{undefined if} & ru \notin R. \end{cases}$$

Then [11, Theorem IV.5.2] states that each $\rho_u \in \mathcal{I}(R/K)$, that ρ is a morphism, and that $0 = R/K$ is an orbit. Our first key result is the following.

THEOREM 4. *If U is a finite inverse semigroup, then every orbit is obtained, up to isomorphism, from an $\mathcal{R}$-class R and a subgroup K, as above.*

Our second key result is the following combinatorial result.

THEOREM 5. *Let K, L be $n \times n$ upper triangular matrices whose entries are non-negative integers and whose diagonal entries are positive integers. Denote the rows of K by $\underset{\sim}{k}_1, \underset{\sim}{k}_2, \ldots, \underset{\sim}{k}_n$ and those of L by $\underset{\sim}{\ell}_1, \underset{\sim}{\ell}_2, \ldots, \underset{\sim}{\ell}_n$. Then there exist positive integers $\alpha_1, \alpha_2, \ldots, \alpha_n$ and*

$\beta_1, \beta_2, \ldots, \beta_n$ *such that*

$$\alpha_1 \underset{\sim}{k}_1 + \alpha_2 \underset{\sim}{k}_2 + \ldots + \alpha_n \underset{\sim}{k}_n = \beta_1 \underset{\sim}{\ell}_1 + \beta_2 \underset{\sim}{\ell}_2 + \ldots + \beta_n \underset{\sim}{\ell}_n .$$

REFERENCES

1. Clifford, A.H. and G.B. Preston, *The Algebraic Theory of Semigroups*, Amer. Math. Soc., Mathematical Surveys 7, I and II (Providence, R.I., 1961 and 1967).

2. Grätzer, G. and H. Lakser, *The structure of pseudo-complemented distributive lattices II: congruence extension and amalgamation*, Trans. Amer. Math. Soc. 156 (1971), 343-357.

3. Hall, T.E., *Free products with amalgamation of inverse semigroups*, J. Algebra 34 (1975), 375-385.

4. Hall, T.E., *Amalgamation and inverse and regular semigroups*, Trans. Amer. Math. Soc. 246 (1978), 395-406.

5. Hall, T.E., *Representation extension and amalgamation for semigroups*, Quart. J. Math. Oxford (2), 29 (1978), 309-334.

6. Hall, T.E., *Amalgamation and inverse and regular semigroups: a brief survey*, Proceedings of a Symposium on Regular Semigroups, Northern Illinois University, DeKalb, 1979.

7. Hall, T.E., *On regular semigroups II: an embedding*, J. Pure Appl. Algebra 40 (1986), 215-228.

8. Hall, T.E. and Mohan S. Putcha, *The potential J-relation and amalgamation bases for finite semigroups*, Proc. Amer. Math. Soc. 95 (1985), 361-364.

9. Howie, J.M., *Semigroup amalgams whose cores are inverse semigroups*, Quart. J. Math. Oxford (2), 26 (1975), 23-45.

10. Howie, J.M., *An introduction to semigroup theory*, London Math. Soc. Monographs 7, Academic Press, 1976.

11. Petrich, Mario, *Inverse Semigroups*, John Wiley and Sons, 1984.

12. Schreier, O., *Die untergruppen der freien gruppen*, Abh. Math. Sem. Univ. Hamburg 5 (1927), 161-183.

RANK PROPERTIES IN SEMIGROUPS OF MAPPINGS

John M. Howie
Mathematical Institute
University of St Andrews
North Haugh
St Andrews, Scotland, KY16 9SS

ABSTRACT. The *rank* of a finite semigroup S is defined as $r(S) = \min\{|A| : \langle A\rangle = S\}$. If S is generated by its set E of idempotents or by its set N of nilpotents, then the idempotent rank ir(S) and the nilpotent rank nr(S) are given by $ir(S) = \min\{|A| : A \subseteq E \text{ and } \langle A\rangle = S\}$ and $nr(S) = \min\{|A| : A \subseteq N \text{ and } \langle A\rangle = S\}$ respectively; these are potentially different from r(S). If $Sing_n$ is the semigroup of all singular self-maps of $\{1,\ldots,n\}$ then $r(Sing_n) = ir(Sing_n) = \frac{1}{2}n(n-1)$. If SP_n is the inverse semigroup of all proper subpermutations of $\{1,\ldots,n\}$ then $r(SP_n) = nr(SP_n) = n + 1$.

The rank r(S) of a semigroup S is defined as

$$\min\{|A| : A \subseteq S, \langle A\rangle = S\}. \qquad (1)$$

It is a simple concept, but surprisingly little attention has been paid to it. If $|S| > \aleph_0$ the concept is uninteresting since $r(S) = |S|$; but for finite or countable semigroups it is not always easy to calculate r(S).

If G is a finite group then

$$r(G) \leqslant \log_2 |G|.$$

[If $G = \langle a_1\rangle$ has rank 1 then $|G| \geqslant 2$. Suppose inductively that if $r(G) = n-1$ then $|G| \geqslant 2^{n-1}$. Then let $G = \langle a_1,\ldots,a_{n-1},a_n\rangle$ have rank n. Since $H = \langle a_1,\ldots,a_{n-1}\rangle$ is a proper subgroup of G we must have $(G:H) \geqslant 2$. Hence, since $r(H) = n-1$,

$$|G| = (G:H)|H| \geqslant 2.2^{n-1} = 2^n. \quad]$$

This bound is best possible: if $G = \mathbb{Z}_2 \times \mathbb{Z}_2 \times \ldots \times \mathbb{Z}_2$ (n factors) then $|G| = 2^n$, $r(G) = n$.

S. M. Goberstein and P. M. Higgins (eds.), Semigroups and Their Applications, 57–60.

By contrast, the best we can say for finite semigroups is the crashingly obvious

$$r(S) \leqslant |S|.$$

This bound is attained, for example, by any finite left-zero semigroup. Along with Emilia Giraldes [2,1] I have investigated semigroups such that $r(S) = |S|$ – which are easy – and also semigroups such that $r(S) = |S| - 1$. Somewhat whimsically we called semigroups of the highest possible rank ($r(S) = |S|$) *royal*; those of next highest rank ($r(S) = |S| - 1$) we called *noble*. This enabled Simon Goberstein, in reviewing the paper, to show his support for American constitutional principles by remarking (what is true) that royal semigroups are highly degenerate. They must be bands; hence they are semilattices of rectangular bands. Moreover each of the rectangular bands is left or right zero, and the semilattice is a chain. Noble semigroups are less easy to describe. But that is not what I want to talk about today.

If a semigroup S is generated by its set E of idempotents it is possible to define the *idempotent rank* of S by

$$ir(S) = \min\{|A| : A \subseteq E \text{ and } \langle A\rangle = S\}.$$

Equally, if S is generated by its set N of nilpotents we may define the *nilpotent rank*

$$nr(S) = \min\{|A| : A \subseteq N \text{ and } \langle A\rangle = S\}.$$

It is clear that when $ir(S)$ [or $nr(S)$] is defined then

$$ir(S) \geqslant r(S) \ [nr(S) \geqslant r(S)].$$

The inequality may be strict. For example let

$$S = M^0[\{e\};\{1,2,3\},\{1,2,3\};P],$$

with

$$P = \begin{bmatrix} e & e & 0 \\ 0 & e & e \\ 0 & 0 & e \end{bmatrix}.$$

Then S has nine non-zero elements, which we may identify with the elements (i,j) of $\{1,2,3\}^2$. Then

$$A = \{(1,1),(2,1),(2,2),(3,2),(3,3)\}$$

(consisting of all non-zero idempotents) is the smallest set of idempotent generators, while

$$B = \{(1,2),(2,3),(3,1)\}$$

is a minimal set of generators. Thus

$$r(S)=3, \quad ir(S)=5.$$

Given a finite set $Z = \{1,\ldots,n\}$, let us denote by T_n the full transformation semigroup on Z. It is well-known that $r(T_n) = 3$: specifically, the symmetric group S_n is generated by two elements (12),(12...n), and we obtain a generating set for S_n by adjoining (e.g.) a single idempotent of defect 1.

Also well-known is the fact that $Sing_n = T_n \backslash S_n$ is idempotent-generated, but nobody seems to have investigated $r(Sing_n)$ or $ir(Sing_n)$. The answer turns out to be quite easy and will appear in a paper written jointly with Gracinda Gomes [4]

First, recall that

$$Sing_n = J_1 \cup \ldots \cup J_{n-1},$$

where J_r ($r = 1,\ldots,n-1$) is the $\mathcal{J}$-class consisting of all α in $Sing_n$ such that $|\mathrm{im}\,\alpha| = r$. It is not hard to see that $\langle J_{n-1}\rangle = Sing_n$ and that $A \subseteq J_{n-1}$ for any generating set A of minimal size.

The structure of the $\mathcal{J}$-class J_{n-1} (since the semigroup is finite, $\mathcal{J} = \mathcal{D}$) is easily determined : there are n $\mathcal{L}$-classes, corresponding to the n possible images $Z\backslash\{i\}$ for an element α in J_{n-1}; and there are $\frac{1}{2}n(n-1)$ $\mathcal{R}$-classes, corresponding to the $\frac{1}{2}n(n-1)$ possible partitions of the form $\{\{a_1,a_2\},\{a_3\},\ldots,\{a_n\}\}$ of Z. A minimal generating set A ($\subseteq J_{n-1}$) must certainly generate the principal factor $J_{n-1}/(J_1 \cup \ldots \cup J_{n-2})$ *qua* completely 0-simple semigroup. Since a completely 0-simple semigroup has the property

$$xy \neq 0 \Rightarrow xy \,\mathcal{R}\, x,$$

every generating set must cover the $\mathcal{R}$-classes and so must contain at least $\frac{1}{2}n(n-1)$ elements.

In [5] a generating set is found for $Sing_n$ consisting of $\frac{1}{2}n(n-1)$ idempotents. Hence we have

THEOREM 1. $r(Sing_n) = ir(Sing_n) = \frac{1}{2}n(n-1)$.

Let us look now at I_n, the symmetric inverse semigroup on $Z = \{1,\ldots,n\}$. Since there is a potential ambiguity in defining rank for an inverse semigroup S let us be clear that by $\langle A\rangle$ we shall here mean the *inverse* subsemigroup generated by A; the rank is then defined by (1).

The following result is probably well-known and is anyway easy.

THEOREM 2. $r(I_n) = 3$.

By analogy with $Sing_n$ we may consider $SP_n = I_n \backslash S_n$, the inverse semigroup of all *proper* subpermutations of $Z = \{1,\ldots,n\}$. From [3] we know that SP_n is nilpotent-generated if and only if n is even.

Then

THEOREM 3. $r(SP_n) = n+1$. *If* n *is even then* $nr(SP_n) = n+1$.

If n is odd then the nilpotents in SP_n generate a proper inverse subsemigroup $SP_n \backslash W$ of SP_n. To describe W, note first that if $\alpha \in SP_n$ is such that $|\mathrm{dom}\,\alpha| = n-1$ then α maps $Z\backslash\{i\}$ onto $Z\backslash\{j\}$ bijectively, where $i,j \in Z$. The element $\bar{\alpha} = \alpha \cup \{(i,j)\}$ of S_n is called the *completion* of α. Then

$$W = \{\alpha \in SP_n : |\mathrm{dom}\,\alpha| = n-1,\ \bar{\alpha} \text{ is an odd permutation}\}.$$

THEOREM 4. If n is odd, then

$$r(SP_n\backslash W) = nr(SP_n\backslash W) = n+1.$$

Perhaps of independent interest is the final theorem, which is used in the proofs of Theorems 3 and 4.

THEOREM 5. If $S = B(G;\{1,\ldots,n\})$ is a Brandt semigroup, then $r(S) = r(G) + n - 1$.

REFERENCES

1. EMILIA GIRALDES, Semigroups of high rank. II.'Doubly noble semigroups', *Proc. Edinburgh Math. Soc.* **28** (1985) 409–417.
2. EMILIA GIRALDES and JOHN M. HOWIE,'Semigroups of high rank', *Proc. Edinburgh Math. Soc.* **28** (1985) 13–34.
3. GRACINDA M. S. GOMES and JOHN M. HOWIE,' Nilpotents in finite inverse semigroups' (submitted).
4. GRACINDA M. S. GOMES and JOHN M. HOWIE,'On the ranks of certain finite semigroups of transformations' (submitted).
5. JOHN M. HOWIE, 'Idempotent generators in finite full transformation semigroups', *Proc. Royal Soc. Edinburgh* A **81** (1978) 317–323.

INVERSE SEMIGROUPS WHOSE LATTICES OF FULL INVERSE SUBSEMIGROUPS ARE MODULAR

K.G. Johnston
Department of Mathematics
College of Charleston
Charleston, SC 29424

and

P.R. Jones
Department of Mathematics, Statistics
and Computer Science
Marguette University
Milwaukee, WI 53233

ABSTRACT. An inverse semigroup is said to be modular if its lattice $\mathcal{LF}(S)$ of full inverse subsemigroups is modular. We show that it is sufficient to study simple inverse semigroups which are not groups. Our main theorem states that such a semigroup S is modular if and only if (I) S is combinatorial, (II) its semilattice E of idempotents is "Archimedean" in S, (III) its maximum group homomorphic image G is locally cyclic and (IV) the poset of idempotents of each $\mathcal{D}$-class of S is either a chain or contains exactly one pair of incomparable elements, each of which is maximal. It is shown that there is exactly one bisimple modular inverse semigroup which is not a group and that is nondistributive.

1. INTRODUCTION

The study of the lattice of subalgebras of an algebra has a long history. In particular, numerous authors have considered the lattice of all subsemigroups of a semigroup (see for example the survey article [9]), the lattice of subgroups of a group (see [10], the lattice of completely simple subsemigroups of a completely simple semigroup ([5]). (These last three can be regarded as the lattice of subalgebras when the appropriate semigroups are treated as algebras with multiplication and an additional unary operation.) The question then arises as to what semigroups have subalgebra lattices satisfying such lattice - theoretic properties as semi-modularity, modularity and distributivity. In [2] it was shown that subalgebra lattices of semilattices satisfy no nontrivial lattice identities. Eršova showed in [1] that an inverse semigroup whose lattice of inverse subsemigroups is semi-modular has a very simple structure.

Full proofs of the new results given below will appear in [6].

S. M. Goberstein and P. M. Higgins (eds.), Semigroups and Their Applications, 61–67.

2. THE LATTICE LF

The second author initiated the study of the lattice $LF(S)$ of full inverse subsemigroups of an inverse semigroup S in [6]. (A full inverse subsemigroup of S is one containing all the idempotents of S.) The lattice $LF(S)$ (or just LF) is a complete sublattice of the lattice of all subsemigroups of S, with E as zero element. Note that if G is a group, then $LF(G) = L(G)$, the lattice of all subgroups of G. An inverse semigroup S is said to be modular [semimodular, distributive] if $LF(S)$ is.

The following result simplifies the study of LF.

RESULT 1. [7]. *For any inverse semigroup* S, $LF(S)$ *is a subdirect product of the lattices of full inverse subsemigroups of its principal factors. Hence* S *is modular* [*semimodular, distributive*] *if and only if each of its principal factors is.*

Thus the problem is reduced to studying LF for 0-simple inverse semigroups. In the special case of completely 0-simple inverse semigroups, those which are modular, semimodular and distributive have been completely described, as follows.

RESULT 2. [7]. *A semimodular* [*modular, distributive*] *completely 0-simple inverse semigroup* S *is either a semimodular* [*modular, distributive*] *group with zero adjoined or is a combinatorial Brandt semigroup* [*with at most three, two nonzero idempotents*].

In the case that S is inverse and 0-simple (but not completely 0-simple), necessary and sufficient conditions were given in [8] for S to be distributive.

3. MODULAR INVERSE SEMIGROUPS

Here we investigate the structure of modular inverse semigroups which are 0-simple but not completely 0-simple. We first show that only simple inverse semigroups need be considered.

PROPOSITION 3. *Let* S *be a modular inverse semigroup which is 0-simple but not completely 0-simple. Put* $S^* = S\setminus\{0\}$. *Then* S *has no zero divisors, whence* S^* *is a simple inverse semigroup with* $LF(S^*) \cong LF(S)$.

DEFINITION. An element x is called strictly right regular if $xx^{-1} > x^{-1}x$, in other words if the inverse subsemigroup generated by x is the bicyclic semigroup. The semilattice of idempotents E of S is said to be Archimedean in S if for any $e, f \in E$ and any strictly right regular element x of R_e, $x^{-n}x^n \leqslant f$ for some positive integer n.

We can now state our main theorem.

MAIN THEOREM. *A simple inverse semigroup* S *(not a group) with semilattice* E *of idempotents is modular if and only if*

(I) S *is combinatorial,*

(II) E *is Archimedean in* S,

(III) *the maximum group homomorphic image of* S *is locally cyclic and*

(IV) *the poset of idempotents of each* $\mathcal{D}$*-class of* S *is either a chain or contains exactly one pair of incomparable elements, each of which is maximal.*

This theorem is remarkable for its similarity to the description of distributive simple inverse semigroups.

THEOREM 4.[8]. *A simple inverse semigroup* S *is distributive if and only if* S *satisfies (I) - (III) above and the poset of idempotents of each* $\mathcal{D}$*-class is a chain.*

We will call S locally distributive if each local submonoid eSe, $e \in E$, is distributive. (This usage differs from that current in lattice theory.) The next result gives another illustration of how close modular inverse semigroups are to being distributive.

PROPOSITION 5. *A simple inverse semigroup* S *is modular if and only if* S *is locally distributive and (IV) holds.*

We now review some terminology and introduce some notation.

The kernel, ker γ, of a congruence γ on an inverse semigroup S is the full inverse subsemigroup $\{x \in S : x\gamma e$ for some $e \in E\}$. Denote by $\sigma = \{(x,y) : ex = ey$ for some $e \in E\}$. The following result shows that if S is modular, so is S/σ.

PROPOSITION 6. *Let* γ *be a congruence on an inverse semigroup* S. *The map* $\Gamma : LF(S) \to LF(S/\gamma)$ *is a join-preserving surjection. When restricted to the principal filter* $[\ker \gamma, S]$, Γ *is a homomorphism upon* $LF(S/\gamma)$. *Hence if* S *is modular so is* S/γ.

Throughout, we shall denote the maximum group homomorphic image S/σ by G. Let $K = \ker \sigma = \{x : ex = e$ for some $e \in E\}$. Then S is E-unitary if $K = E$. Another characterization of K is given in the following theorem.

THEOREM 7. *If* S *is a simple modular inverse semigroup which is not a group, then* $K = \{x : xx^{-1} \parallel x^{-1}x\} \cup E = \{x : x^2 = x^3\}$.

We next elucidate further properties of K and G.

THEOREM 8. *Let* S *be a simple modular inverse semigroup which is not a group. Then*

(i) S *is locally* E*-unitary, and* E*-unitary if and only if distributive;*

(ii) G *is distributive and is isomorphic with a subgroup of the additive group* Q *of rationals;*

(iii) K *is distributive.*

PROOF.

(i) That S is locally E-unitary follows from the fact that local submonoids inherit simplicity, and that simple distributive inverse semigroups are E-unitary [8, Lemma 2.7]. If S is E-unitary then $K = E$. By Theorem 7, xx^{-1} and $x^{-1}x$ are comparable for every $x \in S$, that is, $(E_D, \leqslant)$ is a chain for each $\mathcal{D}$-class D, whence S is distributive by Theorem 4.

(ii) It follows from [11, Proposition 1.12] that a torsion-free modular group is abelian. To see that G is torsion-free, let $x \in S$ and suppose $x^n \in K$ for some $x > 0$, that is, $(x\sigma)^x = 1$ in G. By Theorem 7, $x^n x^{-n} \parallel x^{-n} x^n$. But then, by considering the known structures of monogenic inverse semigroups (see, for instance [9, Chapter IX]), we have that the $n + 1$ $\mathcal{D}$-related idempotents $x^n x^{-n}$, $(x^{-1}x)(x^{n-1}x^{-(n-1)}), \ldots, (x^{-i}x^{i})(x^{n-i}x^{-(n-i)}), \ldots, x^{-n}x^{n}$ form an antichain. By (IV), $n + 1 \leqslant 2$, that is, $n = 1$ and $x\sigma = 1$ in G, as required. Hence G is abelian. A group G is distributive if and only if it is locally cyclic [11, Theorem 1.2], so G is distributive. Thus a torsion-free distributive group is an abelian group of rank 1. It is therefore isomorphic to a subgroup of the additive group Q of rationals ([3]).

(iii) K is completely semisimple and each of its principal factors contains at most two non-zero idempotents, so distributivity follows from Result 2.

We now consider some examples. In view of (IV), a semilattice which is bound to play an important role is that in Figure 1, which we shall denote by Y. It is obtained from the from the ω-chain $C_\omega = \{e_0 > e_1 > e_2 > \ldots\}$ by adjoining two elements f_1 and f_2 covering e_0.

Let $x \in R_{g_1} \cap L_{h_0}$, and put $h_i = x^{-(i+1)} x^{i+1}$, $i \geqslant 1$. Now g_1 and g_2 cover h_0, and the map $e \to x^{-1} ex$ is an isomorphism of Eg_1 onto Eh_0, so $x^{-1}x$ covers $x^{-2}x^2$. By induction, $x^{-i}x^i$ covers $x^{-(i+1)}x^{i+1}$, that is, h_{i-1} covers h_i, $i \geqslant 1$. Let $h \in Eh_0$. Since E is Archimedean in S, there is a least positive ingeger n such that $h_{n-1} = x^{-n}x^n \leqslant h$. But Eh_0 is a chain, so $h_{n-2} > h \geqslant h_{n-1}$, whence $h = h_{n-1}$. Hence $E = \{g_1, g_2\} \cup \{h_n : n \geqslant 0\}$, and is clearly isomorphic with Y.

Since S is combinatorial it is fundamental, and so is isomorphic to a transitive inverse subsemigroup of T_Y ([4, Theorem V.6.4]). But T_Y is combinatorial, so it is its only transitive subsemigroup. Thus $S \cong T_Y$.

We now consider some examples which are not bisimple. Let A be the (full) inverse subsemigroup of T_Y generated by α^2 and β^2 and the set of idempotents. Since T_Y is modular, so is A. Now A has two $\mathcal{D}$-classes, whose idempotents are $E(D_{f_1}) = \{f_1, f_2, e_1, e_3, e_5, \ldots\}$ and $E(D_{e_0}) = \{e_0, e_2, e_4, \ldots\}$. Thus A is simple but not bisimple. In contrast with the distributive case, only the $\mathcal{D}$-class D_{e_0} here is a subsemigroup.

That local distributivity does not imply modularity is clear on consideration of the Munn semigroup of the semilattice obtained from Y by adjoining a further covering element f_3 of e_0.

The strong restrictions placed on the posets of idempotents of $\mathcal{D}$-classes of modular inverse semigroups do not mean that the whole semilattice cannot be quite complicated.

PROPOSITION 10. *Any semilattice can be embedded in the semilattice of idempotents of some simple distributive inverse semigroup.*

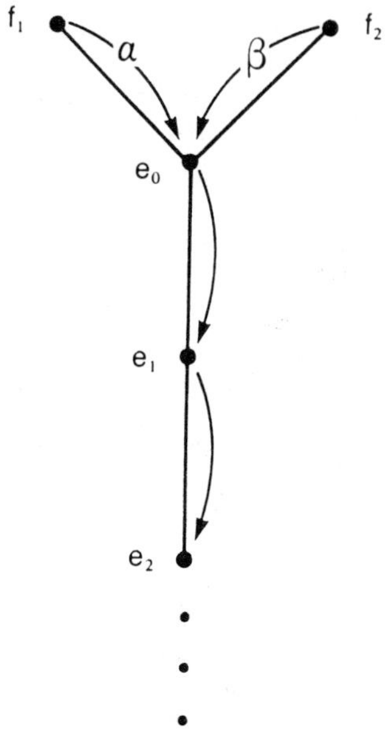

Figure 1

Each principal ideal of Y is isomorphic to C_ω so Y is uniform. The Munn semigroup T_Y is therefore bisimple. (The reader is referred to [4, Chapter V] for details.) Each local submonoid eT_Ye is clearly bicyclic, so T_Y is locally distributive. In view of Proposition 5, T_Y is therefore modular. Note that T_Y is generated by the two σ-related transformations $\alpha : Yf_1 \to Ye_0$ and $\beta : Yf_2 \to Ye_0$ shown in Figure 1, and that also $T_Y/\sigma \cong (Z,+)$.

THEOREM 9. *Let* Y *be the semilattice in Figure 1. The Munn semigroup* T_Y *is the only modular nondistributive bisimple inverse semigroup which is not a group.*

PROOF. Let S be such a semigroup with semilattice E. Since S is nondistributive, E contains exactly two maximal idempotents, g_1 and g_2, say. Put $h_0 = g_1g_2$. By (IV), $E = \{g_1g_2\} \cup Eh_0$, and Eh_0 is a chain.

PROOF. Let X be a semilattice. We may assume $X = X^1$. Form the Bruck-Reilly extension $S = BR(X,\theta)$, where $\theta : X \to \{1\}$. (See [4, V.6] or [9, II.5]). Then S is a simple combinatorial inverse monoid. The semilattice E is isomorphic with the ordinal product of C_ω and X. (See [9, Corollary II.5.13]). Each element of X corresponds to a unique $\mathcal{D}$-class D whose poset of idempotents is isomorphic with C_ω. It is easily verified that E is Archimedean in S and $S/\sigma \cong (Z,+)$. Therefore S is distributive.

By replacing the chain C_ω by the semilattice Y in this construction, a class of simple modular nondistributive inverse semigroups can be constructed. We omit the details.

REFERENCES

1. T. Eršova, "Inverse semigroups with certain types of lattices of inverse subsemigroups", *Ural. Gos. Univ. Mat. Zap.* 7 (1969/70), tetrad' 1, 62-76, (Russian) (MR 43, 6348).

2. R. Freese and J.B. Nation, "Congruence lattices of semilattices", *Pacific J. Math.* 49 (1973), 51-58.

3. L. Fuchs, "*Infinite Abelian Groups*, Vol.I", Academic Press, New York, 1970.

4. J.M. Howie, "*An Introduction to Semigroup Theory*", Academic Press, Londong, 1976.

5. K.G. Johnston, "Subalgebra lattices of completely simple semigroups", *Semigroup Forum* 29 (1984), 109-121.

6. K.G. Johnston and P.R. Jones, "Modular inverse semigroups", *J. Austral. Math. Soc.* (to appear).

7. P.R. Jones, "Semimodular inverse semigroups", *J. London Math. Soc.* 17 (1978), 446-456.

8. P.R. Jones, "Distributive inverse semigroups", *J. London Math. Soc.* 17 (1978), 457-466.

9. M. Petrich, "*Inverse Semigroups*", Wiley, New York, 1984.

10. L.N. Ševrin and A.J. Ovsyannikov, "Semigroups and their subsemigroup lattices", *Semigroup Forum* 27 (1983), 1-154.

11. M. Suzuki, "*Structure of a Group and the Structure of its Lattice of Subgroups*", Springer, Berlin, 1956.

BASIS PROPERTIES, EXCHANGE PROPERTIES AND EMBEDDINGS IN IDEMPOTENT-FREE SEMIGROUPS

Peter R. Jones
Marquette University
Department of Mathematics, Statistics and Computer Science
Milwaukee, WI 53233

ABSTRACT. Two "basis properties" are considered for semigroups and associated algebras. These were introduced by the author to study inverse semigroups and groups, motivated by well-known properties of vector spaces. First these basis properties are studied in the abstract, from the point of view of exchange properties; then those semigroups with the "strong" basis property are determined for many classes of semigroups, including all regular and all periodic semigroups. The main method of proof is to eliminate undesirable types of semigroups by showing that each contains a certain special subsemigroup, for instance the bicyclic semigroup. This prompts the study of analogs of a theorem of Andersen on embeddings of the bicyclic semigroup: such analogs are found for the semigroups $A = \langle a,b \mid a^2b = a\rangle$ and $C = \langle a,b \mid a^2b = a, ab^2 = b\rangle$.

Beginning with a classical theorem of linear algebra, two "basis properties" are considered for semigroups and associated algebras. The <u>basis property</u> (BP) for an algebra requires that any two minimal generating sets ("bases") of a subalgebra have the same cardinality. It turns out that a slightly stronger property, the <u>strong basis property</u> (SBP), is more amenable - it is now required that any two "relative bases" for a subalgebra with respect to one of <u>its</u> subalgebras must have the same cardinality.

In §1 we examine the relationship between basis properties and the classical <u>exchange property</u>. The SBP is equivalent to a certain <u>weak exchange property</u>.

In §2 and §3 we survey old results on basis properties for groups and inverse semigroups. In §4 these properties are studied for semigroups and monoids. The semigroups with SBP are determined for many (possibly all) classes of semigroups, including regular, periodic etc.

The use of the bicyclic semigroup in §4, to eliminate subsemigroups of certain types, prompts investigation in §5 of subsemigroups with analogous properties in the remaining situation where BP is unclear: 0-simple semigroups without nonzero idempotents. This natural question, following the thesis of Andersen (1952), appears to have lain dormant. We show that two semigroups $A = \langle a,b \mid a^2b = a\rangle$

S. M. Goberstein and P. M. Higgins (eds.), Semigroups and Their Applications, 69–82.

and $C = \langle a,b \mid a^2b = a, ab^2 = b\rangle$ play roles, in idempotent-free 0-simple semigroups in which $\mathcal{D}$ is nontrivial, similar to that of the bicyclic semigroup in 0-simple semigroups with primitive idempotents: either C or A or its dual must appear.

When $\mathcal{D}$ is trivial the situation is less clear but we show, at least, that every $\mathcal{D}$-trivial idempotent-free 0-simple semigroup contains a free subsemigroup of countably infinite rank. Combining the results of §§5,6 we deduce that the bicyclic semigroup divides any [0-] simple semigroup which is not completely [0-] simple.

The results of §1 are from [13], those of §§2,3 from [8,9,10], those of §4 from [14] and those of §§5,6 from [11,12].

1. Exchange properties and basis properties

The motivating example for this topic is the classical one of vector spaces over a field: any two linearly independent subsets with the same span (that is, generating the same subspace) have the same cardinality. In universal algebraic terms this translates into what we call the "basis property" (abbreviated henceforth to BP). A basis for a subalgebra B of an algebra A is a minimal generating set for B. (In [2] this is termed an "irredundant basis"). An algebra A is said to have the basis property if

> (BP) *for any subalgebra* B *of* A, *any two bases for* B *have the same cardinality.*

Several comments are in order.

(1) BP is "hereditary", that is, preserved by subalgebras.

(2) Since subalgebra generation is a property of finite character, only finite bases need be treated. Thus the existence or nonexistence of bases for infinitely generated subalgebras is largely irrelevant.

(3) In a standard way (see [2]) vector spaces may be regarded as algebras by adding, for each scalar, a unary operation corresponding to multiplication by that scalar. Thus vector spaces have BP.

(4) Every subalgebra of an algebra A with BP has a well-defined "dimension" (possibly infinite). In contrast with vector spaces, the dimension of a subalgebra may be larger than that of the algebra itself.

(5) Even when an algebra A does not have BP the distribution of possible cardinalities for bases of A may be regular -- see [2, §II.4].

BP for vector spaces is usually derived, implicitly or explicitly, from the "exchange property", which we recall shortly. From this property can also be derived such additional features as extensibility of a basis for a subspace to one for the whole space and monotonicity of the dimension function. (For an exposition in the setting of abstract dependence relations see [4]).

In an algebra A the subalgebra generated by a subset X will be denoted by $\langle X\rangle$. (It will be convenient to allow the empty set to generate the minimum subalgebra of A, if A has no nullary operations). Then A has the <u>exchange property</u> if, for any $X \subseteq A$ and $x, y \in A$,

(EP) *if* $y \in \langle X \cup \{x\}\rangle$ *but* $y \notin \langle X\rangle$ *then* $x \in \langle X \cup \{y\}\rangle$.

Informally, x may be "replaced" by every element y which cannot be expressed in the symbols $X \cup \{x\}$ without using x. In semigroup terms it is clear that for every such y, $J_y \le J_x$. However unless $J_x = J_y$, x cannot also be expressed in terms of y. Thus EP is rare in semigroups. The following two examples illustrate this.

EXAMPLE 1. Let S be the semilattice $\{e,f,0\}$, where e and f are incomparable and $ef = 0$. Then $0 \in \langle\langle e\rangle \cup \{f\}\rangle$ and $0 \notin \langle\langle e\rangle\rangle$ but $f \notin \langle\langle e\rangle \cup \{0\}\rangle$. However BP holds since bases are unique.

EXAMPLE 2. Let S be the semigroup of Figure 1, the ideal extension of the right zero semigroup $K = \{f_0, f_1, \ldots, f_{p-1}\}$ by the group $G = \langle g\rangle$ of order p, determined by $f_i g = f_{i+1}$ (mod p). By similar reasoning, $f_1 \in \langle\langle f_0\rangle \cup \{g\}\rangle$ and $f_1 \notin \langle\langle f_0\rangle\rangle$ but $g \notin \langle\langle f_0\rangle \cup \{f_1\}\rangle$.

The bases of S are precisely the subsets of the form $\{g^k, f_i\}$, $1 \le k < p$, $0 \le i < p$. Clearly the bases of G have the form $\{g^k\}$, $1 \le k < p$, and the sub-bands of S are their own unique bases. Hence, S also has BP. Notice that S has dimension 2 but K has dimension p.

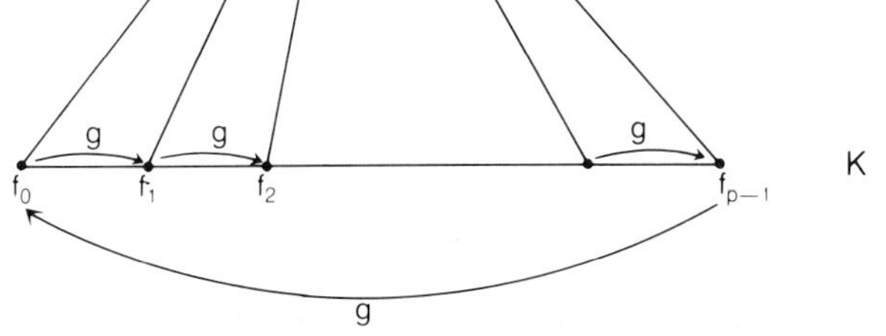

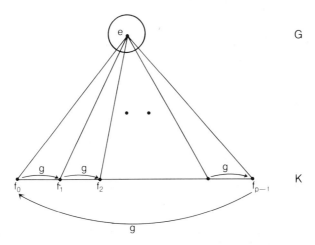

Figure 1. The semigroup in Example 2.

Many generalizations of EP have been proposed. (For a comprehensive survey see [7]). We introduce another which, to our knowledge, is new. As we shall see this arose naturally as an "axiomatic" version of an important basis property, rather than as an empty generalization.

An algebra A has the *weak exchange property* if, for any $X, Y \subseteq A$ and $x \in A$,

(WEP) *if* $\langle Y\rangle = \langle X \cup \{x\}\rangle$ *but* $\langle Y\rangle \neq \langle X\rangle$ *then* $x \in \langle X \cup \{y\}\rangle$ *for some* $y \in Y$.

Informally, again, any "irredundant" generator can be "replaced" by *some* element of any other generating set. (Note that the phrase "but $\langle Y\rangle \neq \langle X\rangle$" may be omitted). That EP implies WEP is obvious; WEP is sufficient to prove BP, moreover something stronger. In fact it arose as an axiomatic version of the following "strong" basis property. If A is an algebra and U, V are subalgebras with $U \subseteq V$ then a U-*basis* for V is a subset X minimal with respect to $\langle U \cup X\rangle = V$. We say A has the *strong basis property* if

(SBP) *for any subalgebras* $U \subseteq V$ *of* A, *any two* U-*bases for* V *have the same cardinality.*

Comments akin to those for BP apply. In particular any subalgebra V of an algebra A has a "relative dimension" with respect to any of *its* subalgebras.

By setting U as the minimum subalgebra of A, it is clear that SBP implies BP. Historically, SBP arose primarily as a technically powerful tool for studying BP (see §3), via the equivalence of (ii) and (iii) in the theorem below, which enables concentration on the "relative dimension one" case. The equivalence of (i) and (ii) answers a question first posed to me by G. B. Preston in 1974.

THEOREM 1.1 ([13]). *The following are equivalent for an algebra* A:

(i) WEP,

(ii) SBP,

(iii) *if* $\langle U \cup \{x\}\rangle = \langle U \cup \{y,z\}\rangle$, *for some subalgebra* U *of* A *and* $x,y,z \in A$, *then either* $x \in \langle U \cup \{y\}\rangle$ *or* $x \in \langle U \cup \{z\}\rangle$.

Each is implied by EP *and implies* BP.

PROOF. The equivalence of (ii) and (iii) is implicit in the proof of [9, Theorem 2.3] (see also [16, §8.5]).

To prove (i) implies (iii), set $X = U$ and $Y = U \cup \{y,z\}$. From WEP it follows that either $x \in \langle X \cup \{y\}\rangle$ or $x \in \langle X \cup \{z\}\rangle$ (or $x \in \langle X\rangle$).

To prove (ii) implies (i), suppose $\langle Y\rangle = \langle X \cup \{x\}\rangle \backslash \langle X\rangle$. Set $U = \langle X\rangle$ and $V = \langle Y\rangle$, so that $\langle U \cup Y\rangle = \langle U \cup \{x\}\rangle$: then $\langle U \cup Y'\rangle = \langle U \cup \{x\}\rangle$ for some *finite* subset Y' of Y. In fact if

$x \notin U$ then $\{x\}$ is a U-basis for V and so by SBP there is an element y of Y for which $\{y\}$ is also a U-basis for V. Thus $x \in \langle U \cup \{y\}\rangle$.

Although the contents of this section have been expressed in the language of subalgebra generation, they may be more generally treated in terms of closure operators on a set. In that form they appear in [13].

2. Basis properties for groups

Since the group $\mathbb{Z}$ of integers has bases $\{1\}$ and $\{2,3\}$ (qua group) it follows that any group with BP is periodic : in fact every element has prime power order.

In an unpublished manuscript [5] Neil Dickson and I described the finite groups with BP or SBP. We will not include this rather technical description here; however it is not too hard to see the following. (For these and further published results see [10]).

THEOREM 2.1 *A finite group with* BP *is solvable. For a finite nilpotent group* BP *and* SBP *are equivalent, and equivalent to being primary, that is, a* p-*group for some prime* p.

That finite p-groups have BP is essentially the "Burnside Basis Theorem" ([19, 7.3.10]). In [10] an example was provided of a finite group with BP but not SBP.

For infinite groups the situation is much less clear. In [10] the second statement of the theorem was generalized to "$\tilde{N}$-groups": groups having the property that whenever one subgroup is maximal in another, it is normal in it. This well-known class of generalized-nilpotent groups will make an unexpected appearance in the next section.

Whether infinite p-groups have BP I do not know.

3. Basis properties for inverse semigroups

My interest in basis properties began with an admittedly offhand question regarding the free monogenic inverse semigroup. In my thesis I proved the following theorem. Throughout this section inverse semigroups are of course being treated as algebras of type $\langle 2,1\rangle$. Recall that "combinatorial" means that $\mathcal{H}$ is trivial, and "completely semisimple" means no two $\mathcal{D}$-related idempotents are comparable.

THEOREM 3.1 ([8;9]). *Every combinatorial, completely semisimple inverse semigroup, in particular every free inverse semigroup, has* SBP. *An inverse semigroup with* BP *is necessarily completely semisimple.*

It is easily seen that any basis for a free inverse semigroup I generates I freely. This of course fixes the cardinality of such a basis. Since its inverse subsemigroups need not be free the same argument cannot be applied there.

Necessity of complete semisimplicity is immediate from an argument which sets the pattern for similar results in the sequel: the bases {1} and {2,3} for $\mathbb{Z}$ may be "pulled back" to bases {a} and {a^2, a^3} for the bicyclic inverse monoid B with presentation (qua inverse semigroup) $\langle a \mid aa^{-1} = 1\rangle$. Then a theorem of Andersen [3, Theorem 2.54] may be applied.

The original proof of Theorem 3.1, which also appears as Corollary VIII.5.8 in the monograph of Petrich [16], is combinatorial in nature. The next theorem, proved two years later, not only describes completely the inverse semigroups with SBP (modulo certain group-theoretic problems) but also illuminates the combinatorial case. Recall that the principal factors of a completely semisimple inverse semigroup are either completely 0-simple inverse semigroups, that is, Brandt semigroups, or groups. Call a Brandt semigroup proper if it is not merely a group with adjoined zero; SBP for groups and groups with zero (qua inverse semigroups) is clearly equivalent to SBP qua groups.

THEOREM 3.2 [10]. *a) An inverse semigroup has* SBP *if an only if each of its principal factors (necessarily Brandt, or a group) has.*

b) A proper Brandt semigroup has SBP *if and only if its maximal subgroups are primary* $\tilde{N}$*-groups.*

The terms "primary" and "$\tilde{N}$" were defined in §2. From the remarks there it follows that if the maximal subgroups of a proper Brandt semigroup are finite it is necessary and sufficient that they be p-groups for some prime p. In particular the combinatorial case is a trivial consequence.

The BP is not as amenable -- in [10] an example is given of a semilattice of groups, each with BP, which does not itself have BP.

One very nice consequence of SBP in an inverse semigroup S is that not only do any two U-bases for an inverse subsemigroup V have the same number of elements but they even have the same number in any given $\mathcal{J}$-class of S. For this and many further results see [8,9,10].

4. Basis properties for semigroups and monoids

My interest in the topic of basis properties was reawakened recently by a paper by J. Doyen [6] in which he proved the following theorem, where generation is with respect to submonoids.

THEOREM 4.1 [6]. *Any periodic R-trivial monoid has* BP. *Any basis for a J-trivial monoid is unique.*

It is natural to ask whether other interesting classes of semigroups and monoids have BP or SBP, and precisely which semigroups do have these properties. In view of the nature of the results in §3, SBP is likely to be more fruitful. For the moment we consider semigroups only.

Again it is true that SBP (and also BP) is inherited by principal factors. Whether SBP for all principal factors implies

SBP for the semigroup itself is not yet clear, though I conjecture it is so. (This would be a consequence of verification of the conjecture in §6).

Thus we may concentrate, for the present, on 0-simple, simple and null semigroups. It is trivially seen that null semigroups have SBP. Further, by adjunction of a zero the "simple" case may be absorbed in the "0-simple" case, upon which we now concentrate.

Let S be a 0-simple semigroup with a nonzero idempotent e, say. For instance whenever S is regular or periodic or, more generally, "eventually regular" (every element has a power which is regular) this is the case. If e is not primitive then S contains a copy of the bicyclic semigroup, by the theorem of Andersen cited in §3. (See also §§5,6). The importance of the following examples is then evident.

EXAMPLES (i) The group $\mathbb{Z}$ of integers, qua semigroup. Two bases are $\{1,-6\}$ and $\{10,15,-6\}$ so $\mathbb{Z}$ does not have BP. Hence any group with BP (for subsemigroups) is periodic. But then all subsemigroups are subgroups, and BP (and SBP) are the same, whichever way the group is regarded. (The choice of $\{1,-6\}$ instead of the more natural $\{1,-1\}$ as a basis has in mind later application).

(ii) The bicyclic monoid B presented, as a monoid, by $B = \langle a,b \mid ab = 1\rangle$. The above bases for $\mathbb{Z}$ may be "pulled back" to bases $\{a,b^6\}$ and $\{a^{10}, a^{15}, b^6\}$ for B, using the equation $a = a^{25}b^{24}$.

Hence if a 0-simple semigroup with a nonzero idempotent has BP it is completely 0-simple. Again, we call such a semigroup proper if it is not completely simple with adjoined zero. (The improper case is then treated with the completely simple one). The completely [0-] simple semigroups with SBP are determined as follows. (Those with BP may also be determined).

PROPOSITION 4.2 [14] (i) *A completely simple semigroup has* SBP *if and only if it is either a left zero or right zero semigroup or is a group (with* SBP *for subgroups).*

(ii) *A proper completely 0-simple semigroup has* SBP *if and only if it is isomorphic with the five-element combinatorial Brandt semigroup* B_5.

Note that many Brandt semigroups have SBP qua inverse semigroups but not qua semigroups.

The proof of the proposition involves a tedious elimination of all the undesirable configurations of nonzero $\mathcal{H}$-classes in the "egg-box" picture. We illustrate how it is shown that if S, completely 0-simple, contains distinct $\mathcal{R}$-related idempotents then it is a right group with adjoined zero. We omit the remaining cases.

So suppose $e^2 = e$, $g^2 = g$ and $e \mathcal{R} g$, $e \neq g$. If S is not as asserted then there is a nonzero idempotent f, with $R_f \neq R_e$. The two essential configurations which can arise are shown in Figure 2 below, where shaded boxes necessarily contain idempotents (and the other may or may not).

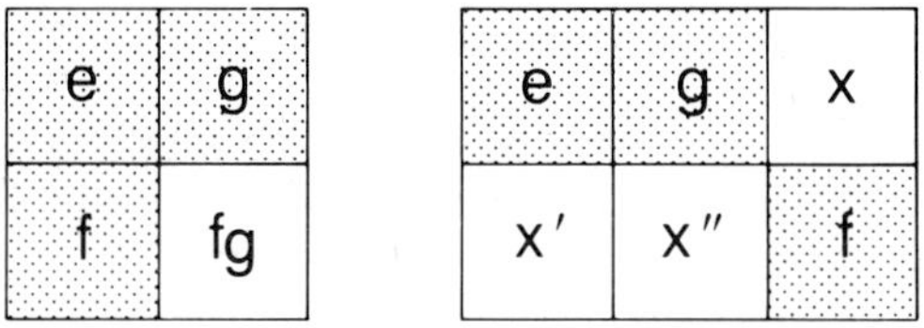

Figure 2. $\mathcal{H}$-classes of S

In the first configuration, both $\{fg\}$ and $\{f,g\}$ are U-bases for $\langle e,f,g\rangle = \langle e,fg\rangle$, with $U = \langle e\rangle$, (since $\langle e,f\rangle \subseteq L_e$ and $\langle e,g\rangle \subseteq R_e$). In the second, where x' and x'' are the inverses of x in their respective $\mathcal{H}$-classes, $\{x''\}$ and $\{g,x'\}$ are U-bases for $\langle e,x,x''\rangle = \langle e,x,g,x'\rangle$, with $U = \langle e,x\rangle$, (since $x'' = x'g$, $g = xx''$, $x' = x''e$ and $\langle e,x,g\rangle \subseteq R_e$, $\langle e,x,x'\rangle \subseteq L_e \cup L_x$).

The remaining type of principal factor not yet considered is the <u>idempotent-free</u> [0-] simple semigroup (containing, that is, no non-zero idempotent). We defer consideration to the next sections.

THEOREM 4.3 [14] *Let* S *be a semigroup with no idempotent-free* [0-] *simple principal factors. Then* S *has* SBP *if and only if each principal factor is either null or has one of the forms in Proposition* 4.1.

As remarked above, the hypotheses apply to all "eventually regular" semigroups, in particular all regular, periodic and, of course, all finite semigroups.

It is clear that in any periodic semigroup with $\mathcal{R}$ trivial, each principal factor is either null or left zero, possibly with adjoined zero. Thus such a semigroup has SBP. If S is a <u>monoid</u>, its submonoids are just the subsemigroups which contain the minimum submonoid $\{1\}$, so SBP holds for S <u>qua</u> monoid. (For a generalization of this idea see [13]). Thus Doyen's result (Theorem 4.1) is an immediate corollary.

In the next sections I will formally conjecture that no idempotent-free [0-] simple semigroup <u>can</u> have BP. Thus I conjecture that in fact the theorem determines <u>all</u> semigroups with SBP.

5. Idempotent-free 0-simple semigroups

The main remaining question in view of §4, is whether an idempotent-free 0-simple semigroup may have BP. I conjecture the answer is no. In this section the conjecture is proved for all such semigroups in which either $\mathcal{R}$ or $\mathcal{L}$ is nontrivial. The method is to look for special subsemigroups which play roles analogous to those of the bicyclic semigroup in Andersen's theorem which, in its complete form, is

THEOREM 5.1 [1;3, Theorem 2.54]. *Any nonzero idempotent in a* [0-] *simple semigroup which is not completely* [0-] *simple is the identity element of a copy of* B.

Owing to its importance, generalizations of this theorem in any direction might be expected to be useful. Whilst various authors have introduced generalizations of the bicyclic semigroup, little effort appears to have been made to find specific generalizations of the theorem itself. My particular need demonstrates how a "curiosity" topic such as basis properties can lead naturally to fundamental questions and their answers.

There is, in the literature, a paucity of examples of the semigroups under discussion. The Baer-Levi semigroups (right simple and right cancellative), Croisot-Teissier semigroups (simple unions of right simple semigroups) and some cancellative examples constructed by Andersen are discussed in [3, Chapter 8].

Let S be an arbitrary 0-simple idempotent-free semigroup. (Our results will also cover the simple case, by adjunction of a zero, but we will state alternative readings when appropriate). First suppose $\mathcal{R}$ is nontrivial. Then clearly $z = zs$ for some $z \in S$, $z \neq 0$. Since S is 0-simple, $s = uzv$ for some $u,v \in S$. Now $z = zuzv$ and $uz = (uz)^2v$. This elementary but key observation is summarized as

LEMMA 5.2. *Any 0-simple idempotent-free semigroup* S *in which* $\mathcal{R}$ *is nontrivial contains elements* x *and* y *satisfying* $x^2y = x$.

Hence, S contains a quotient of the following semigroup

$$A = \langle a,b \mid a^2b = a \rangle .$$

Clearly there are natural homomorphisms

$$A \to B = \langle a,b \mid ab = 1 \rangle \to \mathbb{Z}.$$

As in the previous section, the bases $\{1,-6\}$ and $\{10,15,-6\}$ can be "pulled back" to bases $\{a,b^6\}$ and $\{a^{10}, a^{15}, b^6\}$ of a subsemigroup of A. Thus A itself does not have BP. Obviously we must now study the idempotent-free quotients of A. First the relevant properties of A are summarized. (We omit the multiplication, which is easily determined.) The notation Y^* and Y^+ denotes the free monoid and free semigroup respectively, generated by the set Y.

PROPOSITION 5.3 [12].

(i) *Each element of* $A = \langle a,b \mid a^2b = a \rangle$ *is uniquely expressible as a non-null word of the form*

$$vs^{\ell}a^m, \quad v \in Y^*, \quad \ell \geq 0, \; m \geq 0,$$

where $s = ab$, $Y = \{b_1, b_2, \ldots\}$ *and* $b_n = s^{n-1}b$, $n \geq 1$;

(ii) A *is idempotent-free and right cancellative, whence* $\mathcal{L}$*-trivial;*

(iii) A *has a kernel* K, *comprising all the words not of the form*

b^n; K *is a simple idempotent-free right cancellative semigroup;*
(iv) A *contains the free semigroup* Y^+ *of countably infinite rank.*

THEOREM 5.4 [12]. *The semigroup* A *has a minimum idempotent-free quotient : its "symmetrized" quotient*

$$C = \langle a,b \mid a^2b = a,\ ab^2 = b\rangle.$$

This semigroup C arises naturally when investigating finitely generated simple idempotent-free semigroups. However it has appeared, and its properties have been extensively studied, in papers by Redei [18], Rankin and Reis [17] and Megyesi and Pollák [15]. (For instance, in the last of these papers, C appears as the "greatest" simple combinatorial two-generator semigroup whose one-sided ideals are all principal.) Its relevant properties are also summarized. (These should be compared with Proposition 5.3.)

PROPOSITION 5.5.
(i) *Each element of* $C = \langle a,b \mid a^2b = a, ab^2 = b\rangle$ *is uniquely expressible as a non-null word of the form*

$$b^k s^\ell a^m, \quad k,\ell,m \geq 0,$$

where $s = ab$;
(ii) C *is idempotent-free and simple, neither* R- *nor* L-*trivial;*
(iii) *every proper quotient of* A *contains an idempotent;*
(iv) C *contains no free subsemigroup of rank greater than 1.*

Figure 3 expresses the relationships amongst the semigroups considered; A^d denotes the dual $\langle a,b \mid ab^2 = b\rangle$ of A; D is the pullback of the diagram, presented by $\langle a,b \mid a(ab)^n b = ab,\ n \geq 1\rangle$ (see [12]).

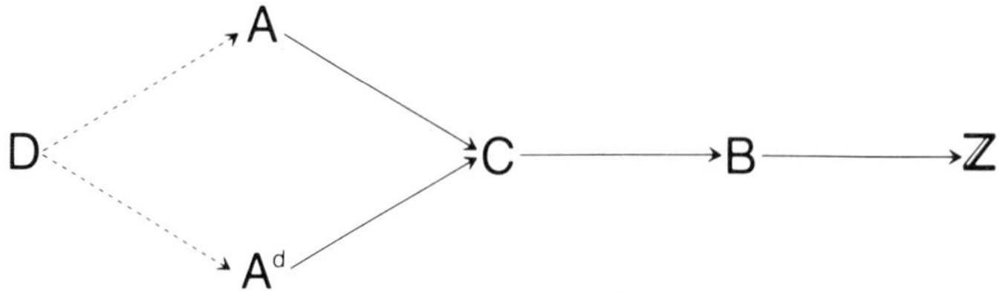

Figure 3. Relationships amongst B and its analogs.

From (ii) and (iv) of the two propositions it is clear that neither A nor C contains an isomorphic copy of the other. In contrast, the essential result for the sequel is the following.

THEOREM 5.6 [12]. *Every proper idempotent-free quotient of* A *satisfies a relation of the form* $s^k v = v$ $(s = ab)$, *where* $v \in Y^+$ *and* $k \geq 1$, *and therefore contains a copy of* C .

The proof of the former statement involves a detailed analysis of the relations satisfied by such a quotient. We show how the latter statement is derived from the former.

Suppose T is such a quotient and k and v are so chosen. Let ℓ be the length of v as a word in Y^+. From the relation $a = as$ in A it is immediate that $ab_n = s$ for all $n \geq 1$, whence $a^\ell v = s$ and $s^{k-1}a^\ell v = s^k$. Then $(s^{k-1}a^\ell)^2 v = (s^{k-1}a^\ell)s^k = s^{k-1}a^\ell$ and $(s^{k-1}a^\ell)v^2 = s^k v = v$ (in T). Hence (the image of) $\langle s^{k-1}a^\ell, v\rangle$ is a subsemigroup of T which is an idempotent-free quotient of C. By Proposition 5.5 (iii), $\langle s^{k-1}a^\ell, v\rangle \cong C$.

Returning to BP, it was noted earlier that it fails in A. Of course, by duality it also fails in A^d and a similar argument shows failure in C. The following theorem is now immediate from Lemma 5.2 and Theorem 5.6 and their duals.

THEOREM 5.7 [14]. *No idempotent-free* 0*-simple semigroup on which either* $\mathcal{R}$ *or* $\mathcal{L}$ *is nontrivial has* BP.

Independently of basis properties the principal result of this section is the analog of Andersen's theorem in this context.

THEOREM 5.8 [12]. *Let* S *be a* 0*-simple idempotent-free semigroup.*

(i) *If* $\mathcal{R}$ *is nontrivial then* S *contains (a copy of)* C *or* A.

(ii) *If* $\mathcal{R}$ *is nontrivial and* $\mathcal{L}$ *is trivial then* S *contains* A *but not* C.

(iii) *If both* $\mathcal{R}$ *and* $\mathcal{L}$ *are nontrivial then* S *contains either* C *or both* A *and its dual* A^d.

When applied to well-known classes of semigroups some striking results are obtained. The first of these is remarkably similar to Andersen's (Theorem 5.1) in character.

COROLLARY 5.9 [12]. *Every simple idempotent-free semigroup with a minimal right ideal, in particular every right simple idempotent-free semigroup, is a union of copies of* A.

Thus every Croisot-Teissier semigroup CT(A, $\mathcal{E}$,p,p) (and every Baer-Levi semigroup of type (p,p) -- see [3, Chapter 8]) is a union of copies of A. That no copy of C appears is due to $\mathcal{L}$-triviality.

On the other hand it is easily seen that on any finitely generated 0-simple idempotent-free semigroup both $\mathcal{L}$ and $\mathcal{R}$ are nontrivial.

COROLLARY 5.10 [12]. *Every finitely generated* 0*-simple idempotent-free semigroup contains either* C *or both* A *and* A^d.

From Theorem 5.5, C is itself an example of such a semigroup, containing neither A nor A^d. However, in [12] examples were found of finitely generated simple idempotent-free semigroups not containing C, and others containing A, A^d and C.

6. The $\mathcal{D}$-trivial case

Finally we consider 0-simple idempotent-free semigroups in which both $\mathcal{R}$ and $\mathcal{L}$ are trivial, that is, which are $\mathcal{D}$-trivial. Cancellative simple idempotent-free semigroups (necessarily $\mathcal{D}$-trivial) were found by Andersen [1; 3, Exercises 2.1.10, 8.4.6], but in general even less seems to be known here. Observe that a [0-] simple $\mathcal{D}$-trivial semigroup with more than one [two] elements is necessarily idempotent-free, by Andersen's theorem.

In view of characterizations of the semigroups embeddable in various special types of simple semigroups it is natural, and relevant, to wonder which semigroups are embeddable in some $\mathcal{D}$-trivial simple semigroup. A clue is found in the proof of Lemma 5.2 : neither of the equations $x = xy$ and $y = xy$ can hold in such a semigroup. These equations play an important role in the exposition in [3, Chapter 8] of the work by Cohn on semigroups embeddable in right simple idempotent-free semigroups, with or without right cancellation. Their place is determined by the following theorem, which utilizes the well-known technique of Šutov [20].

THEOREM 6.1 [11]. *The following are equivalent for a nontrivial semigroup* S :

(i) S *is embeddable in a simple $\mathcal{D}$-trivial semigroup;*

(ii) S *is embeddable in a congruence-free $\mathcal{D}$-trivial semigroup without zero;*

(iii) *neither of the equations* $x = xy$, $y = xy$ *can hold for any* x,y *in* S.

It is natural to again look for analogs of A, B and C. Is there a finite set of "elementary" semigroups, at least one of which must appear in any $\mathcal{D}$-trivial 0-simple semigroup? We cannot answer this question, although there are some natural candidates. One such is the semigroup $_2D$ of Figure 3. Another is the semigroup $U = \langle a,b,c \mid b = ab^2c\rangle$.

Clearly any 0-simple semigroup contains quotients of U, for there must be some element y for which $y^2 \neq 0$, and then elements x and z such that $y = xy^2z$. However, which idempotent-free quotients may appear is not obvious. We observe the following.

PROPOSITION 6.2. *The semigroup* U *does not have* BP .

PROOF. Clearly $U = \langle a,b,c\rangle = \langle a,ab,bc,c\rangle$. Now consider the map of U into the additive group $\mathbb{Z} \oplus \mathbb{Z}$ which takes $a \to (1,0)$, $b \to (-1,-1)$ and $c \to (0,1)$. Since $(1,0) + (-2,-2) + (0,1) = (-1,1)$, this map extends to a homomorphism of U upon $\mathbb{Z} \oplus \mathbb{Z}$. It is now easily seen that <(1,0), (0,-1), (-1,0), (0,1)> is a basis for $\mathbb{Z} \oplus \mathbb{Z}$ (qua semigroup), so <a,ab,bc,c> is a basis for U, contradicting BP.

Although we do not have theorems quite analogous to those of §5 the following interesting result in the same vein holds.

THEOREM 6.3 [12]. *Every $\mathcal{D}$-trivial 0-simple semigroup with more than two elements contains a free subsemigroup of rank greater than* 1.

PROOF. In any quotient of U in which the equation x = xy cannot hold (see Theorem 6.1) it can be shown that the images of ab and a^2b generate a free subsemigroup of rank 2.

Note that this theorem does not extend to the general idempotent-free case, for C itself is a counterexample.

Finally, we may combine the various embeddings to prove the following extension of Andersen's theorem, (where "divides" means "is a homomorphic image of a subsemigroup of").

COROLLARY 6.4 [12]. *The bicyclic semigroup* B *divides every* [0-] *simple semigroup which is not completely* [0-] *simple.*

PROOF. Let S be such a semigroup. If S contains a nonzero idempotent then it actually contains a copy of B. Otherwise either $\mathcal{D}$ is nontrivial, in which case S contains A, A^d or C, each of which has B as quotient, or $\mathcal{D}$ is trivial, in which case B is a quotient of the free subsemigroup found above.

REFERENCES

1. O. Andersen, 'Ein Bericht über die Struktur abstrakter Halbgruppen', Thesis (Staatsexamensarbeit), Hamburg, 1952.
2. S. Burris and H. P. Sankappanavar, A Course in Universal Algebra, Springer-Verlag, New York, 1981.
3. A. H. Clifford and G. B. Preston, 'The Algebraic Theory of Semigroups', Amer. Math. Soc., Providence, Vol. I, 1961, Vol. II, 1967.
4. P. M. Cohn, Universal Algebra, Reidel, Boston, 1981.
5. N. K. Dickson and P. R. Jones, 'Finite groups and the basis property', manuscript, 1976.
6. J. Doyen, 'Equipotence et unicité de systemes générateurs minimaux dans certains monoides', Semigroup Forum **28**, (1984), 341-346.
7. K. Głazek, 'Some old and new problems in independence theory', Colloq. Math. **42** (1979), 127-189.
8. P. R. Jones, 'Inverse Subsemigroups of Free Inverse Semigroups', Thesis, Monash University, 1975.
9. __________, 'A basis theorem for free inverse semigroups', J. Algebra **49** (1977), 172-190.

10. _________, 'Basis properties for inverse semigroups', J. Algebra 50 (1978), 135-152.
11. _________, 'Embedding semigroups in $\mathcal{D}$-trivial semigroups'.
12. _________, 'Analogs of the bicyclic semigroup in simple semigroups without idempotents'.
13. _________, 'Exchange properties and basis properties for closure operators'.
14. _________, 'Basis properties for semigroups'.
15. L. Megyesi and G. Pollák, 'On simple principal ideal semigroups', Studia Sci. Math. Hung. 16 (1981), 437-448.
16. M. Petrich, Inverse Semigroups, Wiley, New York, 1984.
17. S. A. Rankin and C. M. Reis, 'Semigroups with quasi-zeroes', Canadian J. Math. 32 (1980), 511-530.
18. L. Rédei, 'Halbgruppen und Ringe mit Linkseinheiten ohne Linkseinselemente', Acta Math. Acad. Sci. Hungar. 11 (1960), 217-222.
19. W. Scott, Group Theory, Prentice-Hall, New Jersey, 1964.
20. E. G. Šutov, 'Embedding semigroups in simple and complete semigroups', Mat. Sbornik 62 (1963), No. 4; 496-511 (Russian).

LATTICES OF TORSION THEORIES FOR SEMI-AUTOMATA

Dedicated with gratitude to
Prof. Dr. rer. nat. habil. Dr. phil. H.J. Weinert
on the occasion of his 60th birthday

Wilfried Lex
Institut für Informatik
Technische Universität Clausthal
Erzstr. 1
D - 3392 Clausthal-Zellerfeld
FRG

ABSTRACT. The torsion theory for semi-automata in a general sense or acts, as developed in [4], is further investigated. After recalling some of the basic concepts and results of that theory it is proved by means of a lemma on Galois connections in general that the torsion classes, the torsionfree classes, and the torsion theories of semi-automata of an appropriate category form a complete lattice. These lattices are isomorphic to each other or to the dual; they are considered in more detail: it is shown that the abstract classes of irreducible acts form a complete atomistic Boolean sublattice; further a proof is given that the simple abelian groups are characterized as those groups whose lattice of torsion theories for the corresponding group acts is a pentagon.

I. INTRODUCTION

A *semi-automaton*, more precisely an *S-semi-automaton*, $A = {}_SA$, is here understood to be a triple (S,A,δ) where S is a non-empty set, A an arbitrary set, and δ any mapping from SxA into A. This concept, also called *act* or *S-act*, is rather general and covers all the well known applications in semigroups and automata theory.

Even for very small S and A there is an enormous amount of semi-automata ${}_SA$: a set with 3 elements, for example, can act on a set of 4 elements in 16 777 216 ways; so some sort of classification seems to be desirable. The success of the classical torsion theory for modules - one of the most important classes of acts - and the success of many more torsion theories, including that of "abstract relational structures" [2], calls for a torsion theory for semi-automata.

Such a theory was developed by V. Guruswami [3] working with the action preserving mappings and, independently, in another direction, by R. Wiegandt and the author [4], working with finite compositions of

S. M. Goberstein and P. M. Higgins (eds.), Semigroups and Their Applications, 83–90.

embeddings and Rees mappings, s. II. below. For unitary and centered S-acts, where S is a monoid with zero, J.K. Luedeman [5] investigated torsion theories, working with homomorphisms as did Guruswami l.c. In spite of the structural poverty of acts in our sense the theory given in [4] turned out to be surprisingly rich, cf. also [1] and [7].

After a sketch of some results of this theory in II., we will continue the investigation of certain classes of semi-automata by considering the lattices of torsion theories. In III. we shall prove a lemma on Galois connections in general in order to show that the torsion theories of the underlying category form a complete lattice (theorem 1b), which is isomorphic to the lattice of the corresponding torsion classes (th. 1c). The abstract classes of irreducible acts form a complete atomistic Boolean sublattice of that lattice (th. 2). Finally in IV. we shall characterize the non-trivial simple abelian groups as those groups where - under some weak conditions - the torsion theories of the corresponding group acts form a pentagon (th. 3).

II. PRELIMINARIES

First, however, we recall some concepts and results from [4], sometimes slightly modifying the notation used there.

Since an act is essentially a unary algebra it is clear how we have to define a *subact*, a *congruence*, a *homomorphism* etc.

Let $\mathfrak{A} = \{A_\iota \mid \iota \in I\}$ be any set of disjoint subacts $A_\iota = SA_\iota$ $(\iota \in I)$ of an S-act A and $\varkappa(\mathfrak{A})$ the smallest congruence $\varkappa$ of A such that A_ι lies in a class of $\varkappa$ for every $\iota \in I$. We call $\varkappa(\mathfrak{A})$ a *Rees congruence* of A, and write $A \twoheadrightarrow B$ iff there is a Rees congruence ρ with $B \simeq A/\rho$; further we shall write $B \rightarrowtail A$ iff $B \simeq C \leq A$ where $C \leq A$ stands for "C is a subact of A". The preorders $\twoheadrightarrow$ and $\rightarrowtail$ are called *Rees mappings* and *embeddings* respectively.

Though not really using categorical tools we will work in a category $\mathfrak{C}$ where the objects form a *universal* class **U** of S-semi-automata, which means that **U** is *Rees closed*, i.e. closed under Rees mappings, and *hereditary*, i.e. closed under embeddings. The morphisms of $\mathfrak{C}$ are finite compositions of Rees mappings and embeddings. - For the following we will assume that **U** contains at least one non-empty S-act, thus containing already the class **0** of trivial S-acts, i.e. S-acts with at most one element.

A *torsion theory* of $\mathfrak{C}$ is a pair (**T**,**F**) of subclasses of **U** - the torsion class **T** and the *torsionfree class* **F** - with the properties:

$\mathbf{T} \cap \mathbf{F} = \mathbf{0}$,

T is Rees closed,

F is hereditary,

for every act A of $\mathfrak{C}$ there are subacts $A_\iota \in \mathbf{T}$ with $\iota \in I$ and $A/\varkappa(\{A_\iota \mid \iota \in I\}) \in \mathbf{F}$.

In order to use further characterizations of torsion theories we have to recall a few more notations and facts from [4].

The operators Γ and Δ are defined for any subclass $\mathbf{A}$ of $\mathbf{U}$ by

$$\Gamma\ \mathbf{A} \doteqdot \{A\varepsilon\mathbf{U}\ |\forall B\varepsilon\mathbf{A}\ :\ A \nrightarrow B\ \Rightarrow\ B\varepsilon\mathbf{0}\}$$

and

$$\Delta\ \mathbf{A} \doteqdot \{A\varepsilon\mathbf{U}\ |\forall B\varepsilon\mathbf{A}\ :\ B \succ A\ \Rightarrow\ B\varepsilon\mathbf{0}\}$$

respectively, and we get ([4], prop. 7, p.267)

(A) : (Γ,Δ) is a Galois connection between the hereditary and Rees closed classes of $\mathfrak{T}$.

$\mathbf{A}$ ($\subseteq\mathbf{U}$) is said to be *closed under extensions* iff for every $A\varepsilon\mathbf{U}$ and subacts $A_\iota\varepsilon\mathbf{A}$ with $\iota\varepsilon I$ and $A/\varkappa$ $(\{A_\iota \mid \iota\varepsilon I\})$ ε $\mathbf{A}$ one has also $A\varepsilon\mathbf{A}$.

Hereby we achieve ([4], th. 2/3, p. 271/272, or [1], (3)., p. 21)

(B): The following three propositions are equivalent:

1. $(\mathbf{T},\mathbf{F})$ is a torsion theory.
2. a) $\mathbf{T}$ is Rees closed.
   b) $\mathbf{T}$ is *inductive*, i. e. for every ascending chain $A_1 \subsetneq \ldots \subsetneq A_\iota \subsetneq \ldots$, where $A_\iota = SA_\iota$ with $\iota\varepsilon I$ are subacts of an act A ($\varepsilon\mathbf{U}$) and $A_\iota\varepsilon\mathbf{T}$ for $\iota\varepsilon I$, one has $S(\bigcup_{\iota\varepsilon I} A_\iota)\ \varepsilon\ \mathbf{T}$.
   c) $\mathbf{T}$ is closed under extensions.
   d) $\mathbf{0} \subseteq \mathbf{T}$.
   e) $\mathbf{F} = \Delta\mathbf{T}$.
3. a) $\mathbf{F}$ is hereditary.
   b) $\mathbf{F}$ is *coinductive*, i. e. for every descending chain $\rho_1 \geq \ldots \geq \rho_\iota \geq \ldots$, where ρ_ι with $\iota\varepsilon I$ are Rees congruences of an act $A(\varepsilon\mathbf{U})$ and $A/\rho_\iota\ \varepsilon\ \mathbf{F}$ for $\iota\varepsilon I$, one has $A/\bigcap_{\iota\varepsilon I} \rho_\iota\ \varepsilon\ \mathbf{F}$.
   c) $\mathbf{F}$ is closed under extensions.
   d) $\mathbf{0} \subsetneq \mathbf{F}$.
   e) $\mathbf{T} = \Gamma\mathbf{F}$.

III. THE LATTICE OF TORSION THEORIES

As a tool for the following we need a lemma on Galois connections which might be useful also in other contexts and which is very similar to G. Pickert's result that one of the mappings of a Galois connection determines the other [6]:

LEMMA: For a Galois connection (f,g) between $(P,\leq)$ and $(Q,\leq)$ let

$$a \vee b = g(f(a) \wedge f(b)) \qquad (a,b\varepsilon P)$$

and

$$x \vee y = f(g(x) \wedge g(y)) \qquad (x,y\varepsilon Q)$$

where

$$p \wedge q = \inf \{p,q\} \qquad (p,q\ \varepsilon\ P \text{ or } p,q\ \varepsilon\ Q).$$

a) If for $a,b \in P$ ($\neq \emptyset$) and for $x,y \in Q$ the infima always exist then $(gQ,\wedge,\vee)$ and $(fP,\wedge,\vee)$ are lattices.

b) If, in addition to the conditions of a), f and g are surjective or injective, or if f or g is a bijection, then $\mathcal{P} = (P,\wedge,\vee)$ and $\mathcal{Q} = (Q,\wedge,\vee)$ prove to be lattices. Moreover f and g are in this case isomorphisms from $\mathcal{P}$ and $\mathcal{Q}$ onto the dual lattice of $\mathcal{Q}$ and of $\mathcal{P}$, and inverse to g and f, respectively.

c) If infima exist without restriction and the conditions of b) are fulfilled, then the lattices mentioned in b) are complete.

d) Under the assumptions of a)

$$L = \{(a,f(a)) \mid a \in gQ\} = \{(g(x),x) \mid x \in fP\}$$

becomes a sublattice of $(gQ,\wedge,\vee) \otimes (fP,\vee,\wedge)$ by

$$\begin{aligned}(a,x) \wedge (b,y) &= (a \wedge b, x \vee y)\\ (a,x) \vee (b,y) &= (a \vee b, x \wedge y)\end{aligned} \qquad ((a,x),(b,y) \in L).$$

Proof:

a) It suffices to show the assertion for gQ: let $a,b \in P$, $x,y \in Q$ and let $\inf\{a,b\}$, $\inf\{x,y\}$ always exist.

Since a Galois connection (f,g) holds

(1) $$a \leq g(f(a))$$

and

(2) $$g(f(g(x))) = g(x),$$

thus

(3) $$\forall a \in gQ\colon\ g(f(a)) = a.$$

Using $d = g(x) \wedge g(y)$ one gets $d \leq g(x)$ and by (2) so $g(f(d)) \leq g(x)$ and the corresponding for y, hence together

$$g(f(d)) \leq \inf\{g(x), g(y)\} = d$$

and by (1) therefore

$$g(f(g(x) \wedge g(y))) = g(x) \wedge g(y)$$

which proves gQ closed under infima.

Since $a \vee b$ lies by definition in gQ, it remains to show that $a \vee b$ is the supremum of $a,b \in gQ$ in gQ: applying g to $f(a) \wedge f(b) \leq f(a)$ and using (1) one has

$$a \leq g(f(a)) \leq g(f(a) \wedge f(b)) = a \vee b$$

and analogously $b \leq a \vee b$; let be $a,b \leq c \in gQ$, this yields $f(a), f(b) \geq f(c)$, thus $f(a) \wedge f(b) \geq f(c)$ and by (3) therefore

$$a \vee b = g(f(a) \wedge f(b)) \leq g(f(c)) = c.$$

b) If f and g are surjective we have $gQ = P$ and $fP = Q$ and hence by a) the assertion. - For $a,b \in P$ and $f(a) = f(b)$ one immediately obtains $a = b$ from (3), thus the injectivity of f; dually one obtains that of g.

If g is injective (2) gives

(4) $\forall x \in Q: \quad f(g(x)) = x,$

and one gets the injectivity of f dually

$$\forall a \in P: \quad g(f(a)) = a.$$

Thus the injectivity of g and f implies the surjectivity of f and g, which reduces all to the above case.

Since from the injectivity of f, as just proven, the surjectivity of g follows - and dually for g and f -, the case of bijectivity of f or g reduces to that of surjectivity of f and g.

Let f be now a bijection of P onto Q: then (3) gives for $a, b \in P$

(5) $f(a \wedge b) = f(g(f(a)) \wedge g(f(b))) = f(a) \vee f(b)$

and dually one obtains

(5') $f(a \vee b) = f(a) \wedge f(b),$

which proves f to be an isomorphism of $(P,\wedge,\vee)$ onto $(Q,\vee,\wedge)$. Dualization gives the analogue for g.

Application of $\overset{-1}{f}$ to (4) yields the last assertion.

c) The definition of v and the results proven then imply the proposition at once.

d) Because of (3) and its dual one has

$$L = \{(g(f(a)), f(a)) \mid a \in gQ\} \subseteq \{(g(x), x) \mid x \in fP\}$$
$$= \{(g(x), f(g(x))) \mid x \in fP\} \subseteq L$$

and thus the asserted equality.

To recognize $(L,\wedge,\vee)$ as a sublattice of $(gQ,\wedge,\vee) \otimes (fP,\vee,\wedge)$ it is obviously sufficient to verify that L is closed under $\wedge$ and $\vee$, which follows immediately from (5) and (5') respectively.

Next we show

PROPOSITION 1:

a) If subclasses of $\mathbf{U}$ are Rees closed, hereditary, inductive, coinductive or closed under extensions, then also are intersections correspondingly.

b) Arbitrary intersections of torsion classes or torsionfree classes of $\mathfrak{T}$ are again torsion or torsionfree classes respectively.

Proof:

a) The assertions follow at once from the corresponding definitions.

b) Let $\mathbf{C}_\iota$ be torsion classes of $\mathfrak{T}$ for $\iota \in I$ and $\mathbf{D} \leftrightharpoons \bigcap_{\iota \in I} \mathbf{C}_\iota$. By (B) 2.a)

-c) every $\mathbf{C}_\iota$ is Rees closed, inductive and closed under extensions, hence by part a) also $\mathbf{D}$ has these properties; moreover $\mathbf{0} \subseteq \mathbf{C}_\iota$ for $\iota \in I$, thus $\mathbf{0} \subseteq \mathbf{D}$ and therefore $\mathbf{D}$ is a torsion class by (B) 2. - For torsionfree classes one concludes analogously.

Combining this proposition with the lemma and (A) we have immediately

THEOREM 1:

a) The torsion classes and the torsionfree classes of $\mathfrak{T}$ form complete lattices $\mathcal{t} = (\mathcal{T}, \wedge, \vee)$ and $\mathcal{f} = (\mathcal{F}, \wedge, \vee)$ with

$$\mathbf{C} \vee \mathbf{D} \coloneqq \Gamma(\Delta\mathbf{C} \cap \Delta\mathbf{D}) \qquad (\mathbf{C},\mathbf{D} \in \mathcal{T})$$

$$\mathbf{C} \vee \mathbf{D} \coloneqq \Delta(\Gamma\mathbf{C} \cap \Gamma\mathbf{D}) \qquad (\mathbf{C},\mathbf{D} \in \mathcal{F})$$

respectively.

b) The torsion theories of $\mathfrak{T}$ form a complete sublattice $\ell = (\mathcal{L}, \wedge, \vee)$ of $\mathcal{t} \otimes \mathcal{f}'$ - where $\mathcal{f}'$ is the dual lattice to $\mathcal{f}$ - , i.e.

$$\left.\begin{array}{l}(\mathbf{C},\mathbf{D}) \wedge (\mathbf{T},\mathbf{F}) = (\mathbf{C} \wedge \mathbf{T}, \mathbf{D} \vee \mathbf{F}) \\ (\mathbf{C},\mathbf{D}) \vee (\mathbf{T},\mathbf{F}) = (\mathbf{C} \vee \mathbf{T}, \mathbf{D} \wedge \mathbf{F}).\end{array}\right. \qquad ((\mathbf{C},\mathbf{D}),(\mathbf{D},\mathbf{F}) \in \mathcal{L})$$

c) $\mathcal{t} = \mathcal{f}' = \ell$.

The lattice $\mathcal{t}$ of th. 1 is in general highly complicated, but $\mathcal{t}$ possesses pleasant sublattices. In order to describe at least one of these we define an abstract class as an isomorphically closed, $\mathbf{0}$ containing subclass of $\mathbf{U}$. As usual we call an act $A = {}_S A$ irreducible iff it has no proper subacts, i.e. iff A and ${}_S\emptyset$ are the only subacts of A. Herewith we can formulate

THEOREM 2:

The abstract classes of the irreducible acts of $\mathfrak{T}$ form a complete atomistic Boolean sublattice of the lattice $\mathcal{t} = (\mathcal{T}, \wedge, \vee)$ of torsion classes of $\mathfrak{T}$.

Proof:

Let $\mathbf{I}$ be the class of the irreducible acts of $\mathfrak{T}$ and $\mathfrak{A}$ the class of abstract sublasses of $\mathbf{I}$. We have $\mathfrak{A} \subseteq \mathcal{T}$ as already stated in [4], ex. 8, p. 275, and easily verifiable, e.g. by (B) 2.

Let $\mathbf{A},\mathbf{B} \in \mathfrak{A}$, then trivially $\mathbf{A} \cap \mathbf{B} \in \mathfrak{A}$ and $\mathbf{A} \cup \mathbf{B} \in \mathfrak{A}$. Since $\mathbf{A} \vee \mathbf{B}$ is smallest torsion class containing $\mathbf{A} \cup \mathbf{B}$ we obtain $\mathbf{A} \vee \mathbf{B} = \mathbf{A} \cup \mathbf{B}$; the analogue holds also for arbitrary intersection and union, thus $a = (\mathfrak{A}, \cap, \cup)$ is a complete distributive sublattice of $\mathcal{t}$ with $\mathbf{0}$ as zero and $\mathbf{I}$ as unit.

With

$$\tilde{\mathbf{A}} \coloneqq (\mathbf{I} \setminus \mathbf{A}) \cup \mathbf{0} \qquad (\mathbf{A} \in \mathfrak{A})$$

one has $\tilde{\mathbf{A}} \in \mathfrak{A}$ and

$$\mathbf{A} \cap \tilde{\mathbf{A}} = (\mathbf{A} \cap (\mathbf{I} \setminus \mathbf{A})) \cup (\mathbf{A} \cap \mathbf{0}) = \emptyset \cup \mathbf{0} = \mathbf{0},$$

$$A \cup \tilde{A} = A \cup (I \setminus A) = I,$$

hence $\mathfrak{a}$ as a complemented distributive, i.e. as a Boolean, lattice.
The classes containing just one non-trivial act, its isomorphic images, and **0**, are exactly the atoms of $\mathfrak{a}$. Thus every non-trivial element of $\mathfrak{a}$ is a union of atoms and $\mathfrak{a}$ therefore atomistic.

IV. A CHARACTERIZATION OF CERTAIN GROUPS

In order to give the characterization mentioned we have to introduce some further concepts.

Let $A = {}_SA$ be an S-semi-automaton. $a \in A$ is called a *trap* iff $Sa = \{a\}$ and A is said to be a *reset* iff every element of A is a trap. - A is called *fully reducible* iff A is a (disjoint) union of the supports of irreducible acts.

The following result was already given in [4], ex. 5a, 6, p. 274/275 and can easily be verified, e.g. by (B) 2.

PROPOSITION 2:

The resets of $\mathfrak{T}$ form a torsion class, and so do those S-acts of $\mathfrak{T}$ with at least one trap, together with ${}_S\emptyset$.

Using this statemant we shall prove

THEOREM 3:

Let $G = (G, \cdot)$ act as a permutation group on the supports of the G-acts of the universal class $\mathbf{U} = \mathrm{Ob}\,\mathfrak{T}$ which contains the acts with $|G| + 1$ elements: G is a non-trivial simple abelian group if and only if the lattice of its torsion theories of $\mathfrak{T}$ is a pentagon.

Proof:

Let $\mathbb{P}$ denote the set of primes, $\mathfrak{t}$ the lattice of torsion classes of $\mathfrak{T}$ (which by th. 1c is isomorphic to the lattice of torsion theories of $\mathfrak{T}$), and g the cardinality of G.

Let G be a non-trivial simple abelian group, hence $g \in \mathbb{P}$. The acts of $\mathfrak{T}$ are fully reducible, since their supports are disjoint unions of the orbits of G and the irreducible summands are either in **0** or are acts of cardinality g, isomorphic to each other, because the lengths of the orbits divide g.

Let **I** be the class of irreducible acts of $\mathfrak{T}$, **R** that of the resets of $\mathfrak{T}$, and **L** that of those acts in **U** with at least one trap together with ${}_G\emptyset$. Then it is obvious that these are the only non-trivial torsion classes of $\mathfrak{T}$ and $\mathfrak{t}$ is the pentagon:

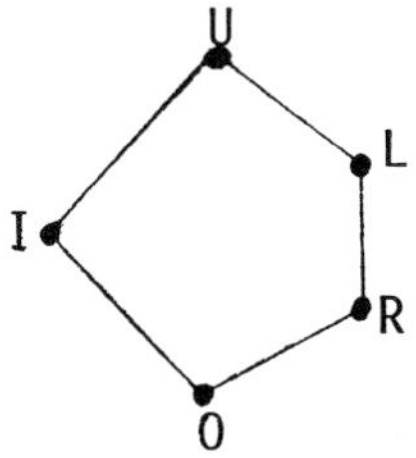

We shall prove the other direction by contraposition. Thus let $g \in \mathbb{P}$. If $g = 1$ one has at once $\mathcal{t}$ as the lattice with 2 elements, since there are only resets.

Let us now assume $g \notin \mathbb{P} \cup \{1\}$. Then there is always a non-trivial proper subgroup $(H,\cdot)$ of G. Therefore **U** contains besides **0** at least 2 torsion classes of irreducible acts, for example that of the acts isomorphic to ${}_GG$ and that of those isomorphic to ${}_GC$ with $C = \{xH \mid x \in G\}$. Indeed ${}_GG$ and ${}_GC$ are not isomorphic, even for $|G| = |C|$, since one has in the first case for every $a \in G$

$$\forall x \in G\,(xa = a \iff x = u)$$

with u as the unit of G, and in the second however

$$\forall x \in G\,(xH = H \iff x \in H),$$

thus no unique determination of x because of $|H| > 1$. Therefore th. 2 implies that the torsion classes of the irreducible acts of $\mathcal{T}$ form a Boolean lattice with at least 4 elements. Thus $\mathcal{t}$ can not be a pentagon.

REFERENCES

1. I. A. Amin, R. Wiegandt: 'Torsion and torsionfree classes of acts.' *Contributions to General Algebra* **2**, Proc. Klagenfurt Conf. 1982. Wien, Stuttgart, 1983; 19-34.

2. E. Fried, R. Wiegandt: 'Abstract relational structures, II (Torsion theory).' *Algebra Universalis*, **15** (1982), 22-39.

3. V. Guruswami: *Torsion theories and localizations for M-sets.* Thesis, Mc Gill University, Montreal 1976.

4. W. Lex, R. Wiegandt: 'Torsion theory for acts.' *Studia Sci. Math. Hungar.*, **16** (1981), 263-280.

5. J. K. Luedeman: 'Torsion theories and semigroups of quotients.' In: K. H. Hofmann, H. Jürgensen, H. J. Weinert (ed.): *Recent Developments in the Algebraic, Analytical, and Topological Theory of Semigroups. Proc., Oberwolfach 1981. Lect. Notes in Math.* **998**. Heidelberg, New York, Tokyo; 1983. 350-373.

6. G. Pickert: 'Bemerkungen über Galois-Verbindungen.' *Arch. Math.*, **III** (1952), 285-289.

7. S. Veldsman, R. Wiegandt: 'On the existence and non-existence of complementary radical and semisimple classes.' *Quaestiones Mathematicae*, **7** (1984), 213-224.

THE COUNTABILITY INDICES OF CERTAIN TRANSFORMATION SEMIGROUPS

K. D. Magill, Jr.
Mathematics Department
106 Diefendorf Hall
SUNY at Buffalo
Buffalo, New York 14214-3093
U.S.A.

ABSTRACT. The countability index $C(S)$ of a semigroup S is defined to be the smallest positive integer N, if it exists, such that every countable subset of S is contained in a subsemigroup with N generators. If no such integer exists, we define $C(S) = \infty$. If S is noncommutative, $C(S) \geq 2$. The rather surprising fact is that it is not all that rare for a full endomorphism semigroup to have countability index two. It has been known for quite a while, for example, that $C(S(I^N)) = 2$ where $S(I^N)$ is the semigroup of all continuous selfmaps of the Euclidean N-cell, I^N. In this paper, we recount the history of the subject and we discuss some recent results concerning the countability index of the endomorphism semigroup of a vector space. We conclude by showing that given any semigroup T, there exists a compact Hausdorff space X such that T can be embedded in $S(X)$ and $C(S(X)) = 2$ where $S(X)$ is the semigroup of all continuous selfmaps of X.

1. HISTORY OF THE SUBJECT

This subject had its genesis in a fundamental paper [5] by J. Schreier and S. Ulam published in 1934. Let $S(I^N)$ denote the semigroup of all continuous selfmaps of the Euclidean N-cell. They established the remarkable fact that there exist five functions in $S(I^N)$ (three continuous selfmaps, a homeomorphism and its inverse) which generate a dense subsemigroup of $S(I^N)$ where the topology on $S(I^N)$ is the compact-open topology (which, in this case, coincides with uniform convergence). In that same year, W. Sierpinski [6] showed that four functions were sufficient for $S(I)$ and one year later, V. Jarnik and V. Knichal [4] showed that, in fact, one could generate a dense subsemigroup of $S(I)$ with only two elements. For one reason or another, nothing further seems to have been done with the problem for the next

S. M. Goberstein and P. M. Higgins (eds.), Semigroups and Their Applications, 91–97.

thirty-four years. Then, in 1969, H. Cook and W. T. Ingram [2] showed that two functions generate a dense subsemigroup of $S(I^N)$ for all N (their result applies to some other spaces as well). Then later, S. Subbiah, not being aware of the results of Cook and Ingram, obtained these results and some others which she published in 1975 [11]. Now, one cannot hope for further improvements because $S(I^N)$ with the compact-open topology is a topological semigroup and if it had a dense subsemigroup generated by one element, it would have to be commutative which, of course, it is not.

It is appropriate to make a few general remarks about how the result is proved for $S(I^N)$. What one does is to show that given any countable subset $\mathcal{F} \subseteq S(I^N)$, there exist two functions which generate a subsemigroup containing $\mathcal{F}$. Then since $S(I^N)$ is separable, one need only choose $\mathcal{F}$ to be a countable dense subset and the result follows. One of the things we found to be of interest in all this is that here is a class of semigroups, each one of which is far from being finitely generated, and yet each countable subset of each one of them is contained in a subsemigroup with only two generators. This led us to the following

<u>DEFINITION (1.1)</u>. The <u>countability index</u> $C(S)$ of a semigroup S is the smallest positive integer n, if such an integer exists, with the property that every countable subset of S is contained in a subsemigroup with n generators. If no such integer exists, we define $C(S) = \infty$.

We will be dealing in this paper with various "natural" transformation semigroups which are noncommutative and so the countability index of such a semigroup must be at least two. The rather surprising fact is that there are a considerable number of instances where the countability index is, in fact, two. In Section 2, we discuss a few results we have obtained recently. We will not give proofs since these results, complete with proofs, will be published elsewhere [10]. Some new results, complete with proofs are presented in Section 3.

2. SOME RECENT RESULTS

The theorems we state in this section are taken from [10]. For proofs, further details and related results, one should consult that paper. First of all, it would be good to have useful necessary and sufficient conditions for a transformation semigroup to have countability index two. We don't have any. We do have a simple necessary condition which is rather easy to state so we'll do so. We also have a set of sufficient conditions which are rather complicated so we'll not state them. For the necessary condition, we need a few preliminary definitions.

<u>DEFINITION (2.1)</u>. Let $T(X)$ be any subsemigroup of $\mathcal{T}_X$, the semi-

group of all selfmaps of the nonempty set X , which satisfies the following conditions.

(2.1.1) T(X) contains the identity map i_X .

(2.1.2) T(X) contains a map f which is not injective.

(2.1.3) T(X) contains a map g which is not surjective

(2.1.4) Either f is surjective or g is injective or T(X) contains an infinite number of bijections.

DEFINITION (2.2). Let $A, B \subseteq X$. Then A and B are T-*isomorphic* if there exist $h, k \in T(X)$ such that $h[A] \subseteq B$, $k[B] \subseteq A$, $k \circ h|A = i_A$ and $h \circ k|B = i_B$.

DEFINITION (2.3). $A \subseteq X$ is a T-*retract* of X if it is the range of an idempotent of T(X) .

THEOREM (2.4). *Suppose* $C(T(X)) = 2$. *Then* X *is* T-*isomorphic to a proper* T-*retract of itself*.

S(X) (the semigroup of all continuous selfmaps of the topological space X) will satisfy the conditions of Definition (2.1) for most spaces X . Nevertheless, there are spaces whose semigroups do not satisfy those conditions. For example, J. de Groot [3] proved the existence (the axiom of choice is used freely) of 2^c mutually non-homeomorphic 1-dimensional, connected subspaces of the Euclidean plane so that for any such space X , S(X) consists of the constant maps together with the identity map. H. Cook [1] also produced a metric continuum with this property. None of these semigroups (in fact, they are all isomorphic) satisfy the conditions of Definition (2.1). Actually, one can easily verify that $C(S(X)) = \infty$ for these semigroups. In spite of these examples, Theorem (2.4) does apply to many S(X) . For example, Brouwer's well known theorem on the invariance of the domain tells us that no Euclidean N-space R^N is homeomorphic to a proper closed subspace of itself. Since, for S(X) , a T-isomorphism is a homeomorphism (though not always conversely) and a T-retract is just a retract in the usual sense, which are closed in Hausdorff spaces, it follows from Brouwer's theorem and Theorem (2.4) that $C(S(R^N)) \geq 3$. In fact, it is known that $C(S(R^N)) \geq 4$ [8] . It is an open problem to determine $C(S(R^N))$.

If we take V to be a vector space over an infinite field F , then End V , the endomorphism semigroup of V satisfies the conditions of Definition (2.1) if $\dim V \geq 1$. Here, two subspaces are T-isomorphic if and only if they are isomorphic in the usual sense and the T-retracts are just the subspaces of V . Since V can only be isomorphic to a proper subspace if it is infinite dimensional, it follows from Theorem (2.4) that for such a vector space V , if C(End V) = 2 , then V is infinite dimensional. We were able to apply our sufficient conditions (given in [10]) to get the converse and all this resulted in the following

THEOREM (2.5). *Let V be a vector space over a field F. Then C(End V) = 2 if and only if either*

(2.5.1) *V is infinite dimensional*

or

(2.5.2) *dim V = 1 and F is finite.*

This characterizes those nontrivial vector spaces V where C(End V) is as small as possible. The next result characterizes those V where C(End V) is as large as possible.

THEOREM (2.6). *Let V be a vector space over a field F. Then C(End V) = ∞ if and only if F is infinite, V is finite dimensional and dim V $\geq$ 1.*

3. SOME NEW RESULTS

In this section, we consider only semigroups of the type S(X) , the semigroup of all continuous selfmaps of the topological space X . We have already noted that $C(S(R^N)) \geq 4$ where R^N is the Euclidean N-space and it is also known that $C(S(S^N)) = \infty$ where S^N is the Euclidean N-sphere [9]. That is, $S(S^N)$ contains countably infinite subsets which are contained in no finitely generated subsemigroups of $S(S^N)$. The question we ask here is this. Is it possible to embed $S(S^N)$ (or $S(R^N)$) algebraically into some S(X) for which C(S(X)) = 2 ? If no restrictions are placed on the space, it follows quickly that the answer is yes. Let $X = S^N$ but endow X with the discrete topology. Then $S(S^N)$ is a subsemigroup of S(X) and it is well known that C(S(X)) = 2 for infinite discrete X [2], [11]. But suppose we look for less trivial spaces. We might, for example ask if X can be taken to be compact. As the next result shows, the answer in this case is also yes.

THEOREM (3.1). *Let T be any semigroup whatsoever. Then there exists a compact, Hausdorff space X such that T can be embedded in S(X) and C(S(X)) = 2.*

It will be convenient to first prove the following

THEOREM (3.2). *Let* X *be any infinite discrete space. Then* $C(S(\beta X)) = 2$ *where, as is customary,* βX *denotes the Stone-Cech compactification of* X .

Proof. The proof relies heavily on Theorem (2.6) of [10]. Decompose X into a countably infinite collection $\{X_n\}_{n=1}^{\infty}$ of mutually disjoint subsets all equipotent with X itself. Then $cl_{\beta X}X_n = \beta X_n$ for each n . Let $A_n = \beta X_n$ and $\{A_n\}_{n=1}^{\infty}$ is a mutually disjoint collection of subspaces of βX , each homeomorphic to βX . Next, decompose X_1 into six mutually disjoint subsets $\{Y_j\}_{j=1}^{6}$ all equipotent with X_1 (and hence with X) and let $B_j = cl_{\beta X}Y_j$, $1 \leq j \leq 6$. We note that $A_1 = \cup\{B_j\}_{j=1}^{6}$. Next, let g_1 be any bijection from X onto X_1 such that g_1 maps $\cup\{X_n\}_{n=2}^{\infty}$ bijectively onto Y_1 , Y_j bijectively onto Y_{j+1} for $1 \leq j \leq 4$ and $Y_5 \cup Y_6$ bijectively onto Y_6 . Then g_1 has a continuous extension $\hat{g}_1$ which maps βX to A_1 and it is easily verified that, in fact, g_1 is a homeomorphism from βX onto A_1 . Now, for $n > 1$, let h_n be any bijection from X_n onto X and let $\hat{h}_n$ be its continuous extension from $A_n = \beta X_n$ to βX . Define $h_1 = \hat{g}_1^{-1}$ and it follows that each $\hat{h}_n$ is a homeomorphism from A_n onto βX .

Now, let $\mathcal{A}$ consist of all nonempty subspaces of βX and for $A,B \in \mathcal{A}$, let $\mathrm{Hom}(A,B)$ consist of all continuous functions from A to B . The pair $(\mathcal{A},\mathcal{M})$ where $\mathcal{M} = \{\mathrm{Hom}(A,B): (A,B) \in \mathcal{A} \times \mathcal{A}\}$ is a Δ_1-structure, as described in Definition (2.5) of [10], whose semigroup is $S(\beta X)$. In particular, the sets $\{A_n\}_{n=1}^{\infty}$ we have just defined satisfy (2.5.3) where K is taken to be the empty set. Now we want to show that the hypothesis of Theorem (2.6) of [10] is satisfied. In fact, condition (2.6.1) is already satisfied by the function $\hat{g}_1$ defined previously. Define $g_2(x) = g_1^{-1}$ for $x \in X_1$ and extend to a selfmap of X in any manner whatsoever. Let $\hat{g}_2$ be the extension of g_2 to a continuous selfmap of βX . Since $\hat{g}_2$ and h_1 agree on X_1 , it follows that they agree on $A_1 = cl_{\beta X}X_1$ so that (2.6.2) is satisfied. For $n > 1$ and $x \in X_n$, define $g_3(x) = h_{n-1}^{-1}(h_n(x))$ and extend in any manner over X_1. Let $\hat{g}_3$ be the extension of g_3 to a continuous selfmap of βX. It is immediate that $\hat{g}_3|A_n = \hat{h}_{n-1}^{-1} \circ \hat{h}_n$ for $n > 1$ and (2.6.3) is satisfied. For each n and $x \in A_n$, define $g_4(x) = h_{n+1}^{-1}(h_n(x))$ and let $\hat{g}_4$ be its extension to a continuous selfmap of βX. Since g_4 is a bijection from X onto $\cup\{X_n\}_{n=2}^{\infty}$, $\hat{g}_4$ is a homeomorphism from βX onto $cl_{\beta X}[\cup\{X_n\}_{n=2}^{\infty}] = \beta[\cup\{X_n\}_{n=2}^{\infty}] = \beta X - A_1$. Evidently, $\hat{g}_4|A_n = \hat{h}_{n+1}^{-1} \circ \hat{h}_n$ and (2.6.4) is satisfied. Next, suppose $k_n \in Hom(A_n, A_n)$ for $n \geq 1$ and for each n and each $x \in X_n$, define $g_5(x) = k_n(x)$. Let $\hat{g}_5$ be the extension of g_5 to a continuous selfmap of βX. Since $\hat{g}_5$ and k_n agree on X_n, they must also agree on $A_n = cl_{\beta X}X_n$ for each n so that (2.6.5) is satisfied.

Finally, let $H = \beta X - A_1$, let $E_i = \hat{g}_1^i[H]$ for $1 \leq i \leq 5$ and $E_6 = g_1^6[\beta X] = cl_{\beta X}Y_6$. It readily follows that $E_i \cap E_j = \phi$ for $i \neq j$ and $A_1 = \cup\{E_j\}_{j=1}^{6}$. Now suppose $k_i \in Hom(E_i, \beta X)$ for $1 \leq i \leq 6$. Define $g_6(x) = k_i(x)$ for $x \in Y_i$, $1 \leq i \leq 6$ and extend in any way whatsoever to a map from X into βX. Let $\hat{g}_6$ be the extension of g_6 to a continuous selfmap of βX. Since $\hat{g}_6|E_i = k_i$ for $1 \leq i \leq 6$, condition (2.6.6) is also satisfied. It now follows from Theorem (2.6) of [10] that $C(S(\beta X)) = 2$ and the proof is complete.

Now we are in a position to complete the

<u>Proof of Theorem (3.1)</u>. Let any semigroup T be given. It is well known that T can be algebraically embedded in $S(X)$ where X is discrete and has sufficiently high cardinality (choose it to be infinite in any case). The mapping which sends $f \in S(X)$ into its continuous extension $\hat{f} \in S(\beta X)$ is easily verified to be a monomorphism. Thus, $S(\beta X)$ contains a copy of T and $C(S(\beta X)) = 2$ by Theorem (3.2).

<u>REMARK</u>: In a certain sense, no proof of Theorem (3.1) can altogether avoid the Stone-Čech compactification of an infinite discrete space. To be more specific, we proved in [7] that if X is arcwise connected, normal and not sequentially compact and $S(X)$ embeds into $S(Y)$ where

Y is compact and Hausdorff, then Y must necessarily contain a copy of βN, the Stone-Čech compactification of the natural numbers.

REFERENCES

1. Cook, H., 'Continua which admit only the identity mapping onto nondegenerate subcontinua', Fund. Math. 60 (1967) 241-249.
2. ________ and W.T. Ingram, 'Obtaining AR-like continua as inverse limits with only two bonding maps', Glasnik Math. 4, Ser. III (1969) 309-312.
3. de Groot, J., 'Groups represented by homeomorphism groups, I', Math. Ann. 138 (1959) 80-112.
4. Jarnik, V. and V. Knichaal,'Sur l'approximation des fonctions continues par les superpositions de deux fonctions', Fund. Math. 24 (1935) 206-208.
5. Schreier, J. and S. Ulam, 'Über topologische Abbildungen der euclidschen Sphäre'. Fund. Math. 23 (1934) 102-118.
6. Sierpinski, W., 'Sur l'approximation des fonctions continues par les superpositions de quarte fonctions', Fund. Math. 23 (1934) 119-120.
7. Magill, Jr., K. D., 'Embedding S(X) into S(Y) when Y is compact and X is not', Semigroup Forum 12 (1976) 347-366.
8. ________, 'Recent results and open problems in semigroups of continuous selfmaps', Uspekhi Mat. Nauk (35) 3 (1980) 78-82; Russian Math. Surv. (35) 3 (1980) 91-97.
9. ________, 'Some open problems and directions for further research in semigroups of continuous selfmaps', Univ. alg. and app., Banach Center Pub., PWN-Polish Sci. Pub., Warsaw, 9 (1982) 439-454.
10. ________, 'The countability index of the endomorphism semigroup of a vector space', (to appear).
11. Subbiah, S., 'Some finitely generated subsemigroups of S(X) ', Fund. Math., 87 (1975) 221-231.

SOME DECISION PROBLEMS FOR INVERSE MONOID PRESENTATIONS

Stuart W. Margolis†, *John C. Meakin*†, *Joseph B. Stephen*†
Department of Computer Science
Department of Mathematics and Statistics
University of Nebraska-Lincoln
Lincoln, Nebraska,68588,U.S.A.

ABSTRACT This paper surveys some of the authors' recent and ongoing work aimed at developing a theory of presentations of inverse monoids analogous to the theory of generators and relations for groups. We regard inverse monoids as a class of algebras of type <2,1,0> and study presentations of inverse monoids from this point of view. The paper is concerned with two basic decision problems for inverse monoid presentations: the word problem and the E-unitary problem. We develop the general construction of a birooted word graph associated with an inverse monoid presentation and show how it can be used as a basic tool in the study of the word problem. We indicate several cases in which the word problem can be solved using these techniques. We study the E-unitary problem for inverse monoids of the form $M=Inv<X|w=1>$ where w is in the free inverse monoid on X. We show how the Lyndon diagrams of combinatorial group theory may be used to analyze the problem and we study several examples and special cases in detail.

1. INTRODUCTION

We consider inverse monoids as a class of algebras equipped with the binary operation of multiplication, the unary operation $a \to a^{-1}$ and the nullary operation of selecting the identity 1 of the monoid. From this point of view, inverse monoids form a variety of algebras of type <2,1,0> subject to the usual laws: $(xy)z = x(yz)$, $x1 = 1x = x$, $xx^{-1}x = x$, $(x^{-1})^{-1} = x$, $(xy)^{-1} = y^{-1}x^{-1}$ and $(xx^{-1})(yy^{-1}) = (yy^{-1})(xx^{-1})$.

We refer the reader to Petrich [14] for details, notations and results concerning inverse semigroups and monoids. In particular, free inverse monoids exist and their structure is discussed in detail in [14, Chapter VIII]. Throughout this paper, X will denote a nonempty set, $X^{-1} = \{x^{-1}|x\epsilon X\}$ a set disjoint from and in one to one correspondence with X, $(X \cup X^{-1})^*$ the free monoid with (obvious) involution on X, ρ the Vagner congruence on $(X \cup X^{-1})^*$ and $FIM(X) \simeq (X \cup X^{-1})^*/\rho$ the free inverse monoid on X.

A presentation of an inverse monoid is a pair $P=(X;R)$ where R is a binary relation on FIM(X). If $R=\{(u_i,v_i)|u_i,v_i \epsilon FIM(X), i\epsilon I\}$, the inverse monoid presented by generators X and relations R is the quotient of FIM(X) by the congruence τ generated by R. Equivalently, we may regard u_i,v_i as elements of $(X \cup X^{-1})^*$ and consider the congruence θ generated by $\rho \cup R$, in which case $M \simeq (X \cup X^{-1})^*/\theta$. We

† Reasearch supported by NSF Grant No. DMS 8503010

S. M. Goberstein and P. M. Higgins (eds.), Semigroups and Their Applications, 99–110.

shall denote the monoid M by $\mathrm{Inv}<X|R>$ or $\mathrm{Inv}<X|\ u_i=v_i,\ i\epsilon I>$.

The free group on X will be denoted by FG(X). The group presented by generators X and relations R will be denoted by $\mathrm{Gp}<X|R>$ and if the u_i and v_i are positive words (i.e. $u_i, v_i \epsilon X^*$) then the monoid presented by X and R will be denoted by $\mathrm{Mon}<X|R>$. The following Lemma is immediate from universal considerations.

LEMMA 1. *The group* $G=\mathrm{Gp}<X|R>$ *is the maximum group homomorphic image of the inverse monoid* $M=\mathrm{Inv}<X|R>$.

Some familiar examples of inverse monoid presentations are the following: FIM(X) is $\mathrm{Inv}<X|\emptyset>$ (the inverse monoid presented by generators X and no relations); the inverse monoid $\mathrm{Inv}<\{a\}|aa^{-1}=1>$ is the bicyclic monoid; the inverse monoid $\mathrm{Inv}<X|\prod_{x\epsilon X}(xx^{-1})(x^{-1}x)=1>$ is the free group on X; the inverse monoid $\mathrm{Inv}<X|xx^{-1}=1, xy^{-1}=0,\ x,y\epsilon X, x\neq y>$ is the polycyclic monoid (see [13]); the inverse monoids $Mc_n=\mathrm{Inv}<\{x_1,\ldots,x_n\}|\ x_ix_j=x_jx_i\ ,1\leq i<j\leq n>$ were considered by McAlister and McFadden [11]-they are of course not commutative since, for example, $x_ix_i^{-1}\neq x_i^{-1}x_i$.

In this paper we provide some information concerning two of the many obvious decision problems that arise when considering inverse monoid presentations. The first of these is the *word problem* : given an inverse monoid presentation $M=\mathrm{Inv}<X|R>$, find an algorithm that will decide for any two words $u,v\epsilon$FIM(X) (or $(X\cup X^{-1})^*$) whether $u\tau v$ or not (that is whether u=v in M). We provide some general techniques for solving this kind of problem in section 2 and illustrate the techniques with a few examples. The second decision problem that we consider is the *E-unitary problem:* what conditions on R guarantee that the inverse monoid $M=\mathrm{Inv}<X|R>$ is E-unitary? See [14] for the concept and importance of E-unitary inverse monoids. In section 3 we provide some information about this problem in the one relator case, that is when $M=\mathrm{Inv}<X|w=1>$ for some $w\epsilon(X\cup X^{-1})^*$.

2. WORD GRAPHS AND THE WORD PROBLEM

Most of the concepts and results of this section are due to Stephen [18], to which we refer for complete details, proofs and additional examples. A *word graph* Γ over X is a connected graph with edges labelled by elements of $X\cup X^{-1}$ such that if

$$(p)\xrightarrow{\ x\ }(q)\ .$$

is an edge of Γ then so is

$$(q)\xrightarrow{\ x^{-1}\ }(p)$$

We normally just indicate the edge

$$(p)\xrightarrow{\ x\ }(q)$$

in the graph Γ. A *birooted word graph* is a triple $A=(\alpha,\Gamma,\omega)$ where Γ is a word graph and α and ω are vertices of Γ. It is convenient to regard A as an automaton over $X\cup X^{-1}$. If $A=(\alpha,\Gamma,\omega)$ is a birooted word graph the *language of* A is given by $L(A)=\{w\epsilon(X\cup X^{-1})^*|\alpha w=\omega\}$. That is, $L(A)$ is the set of words that label paths from α to ω in Γ. For basic definitions and concepts concerning the theory of automata see [3] or [4].

Recall that an automaton A is *deterministic* if the following condition is satisfied. If

$$(p) \xrightarrow{x} (q_1) \quad \text{and} \quad (p) \xrightarrow{x} (q_2)$$

are edges in A for some vertices p, q_1, q_2 and some $x \epsilon X \cup X^{-1}$, then $q_1 = q_2$. Note that if the automaton A of a birooted word graph is deterministic, then it follows that if

$$(p_1) \xrightarrow{x} (q) \quad \text{and} \quad (p_2) \xrightarrow{x} (q)$$

are edges in A for some vertices p_1, p_2, q and some $x \epsilon X \cup X^{-1}$ then $p_1 = p_2$. Thus A is an *inverse automaton*; that is every $x \epsilon\ X \cup X^{-1}$ induces a partial one-to-one map on $V(\Gamma)$ and x^{-1} induces the partial function inverse to that of x. From [15], it follows that if A is deterministic, then A is a minimal automaton. Hence the relation $A \leq B$ if and only if $L(B) \subseteq L(A)$ is a partial order on the set of deterministic birooted word graphs.

Now let $M = \text{Inv}<X|u_i = v_i,\ i \epsilon I>$. We define two operations on birooted word graphs relative to the presentation $P = (X;\{(u_i,v_i),\ i \epsilon I\})$.

(1) **P-Expansions**

Let (α,Γ,ω) be a birooted word graph. If $u_i = v_i$ is a relation in P, and A has a path

$$(p) \xrightarrow{u_i} (q) \quad \text{and} \quad \text{no} \quad \text{path } (p) \xrightarrow{v_i} (q)$$

then we "sew" the path v_i onto A by adding new vertices and edges: thus, if $v_i = x_1 \cdots x_n$ for some $x_j \epsilon X \cup X^{-1}$, then we add new vertices $p_1 \cdots p_{n-1}$ to $V(\Gamma)$ and the new path

$$(p) \xrightarrow{x_1} (p_1) \xrightarrow{x_2} (p_2) \longrightarrow \cdots (p_{n-1}) \xrightarrow{x_n} (q) \quad \text{to } \Gamma.$$

The resulting birooted word graph has n-1 new vertices and n new edges and is said to be obtained from A by a P-expansion.

(2) **P-Reductions**

Let $A = (\alpha,\Gamma,\omega)$ be a birooted word graph. If A has two directed edges with a common initial vertex p and the same label

$$(p) \xrightarrow{x} (q_1) \quad \text{and} \quad (p) \xrightarrow{x} (q_2)$$

for some $x \epsilon X \cup X^{-1}$, we form a new graph by identifying the vertices q_1 and q_2 and the edges (p,x,q_1) and (p,x,q_2). The resulting graph has one fewer edge and one fewer vertex than A and is said to be obtained from A by a P-reduction.

A *P-production* is a P-expansion or a P-reduction. If $(\alpha_1,\Gamma_1,\omega_1)$ is obtained from (α,Γ,ω) by a P-production, we write $(\alpha,\Gamma,\omega) => (\alpha_1,\Gamma_1,\omega_1)$: if $(\alpha_n,\Gamma_n,\omega_n)$ is obtained from (α,Γ,ω) by the sequence of P-productions $(\alpha,\Gamma,\omega) => (\alpha_1,\Gamma_1,\omega_1) => \cdots (\alpha_n,\Gamma_n,\omega_n)$, then we write $(\alpha,\Gamma,\omega) =>^* (\alpha_n,\Gamma_n,\omega_n)$. The following result makes birooted word graphs a useful tool in the study of the word problem for inverse monoid presentations.

LEMMA 2. (The Confluence Lemma, Stephen [18]) *Let P be a presentation of an inverse monoid and let* $A=(\alpha,\Gamma,\omega)$ *be a birooted word graph. If there are P-productions* $(\alpha,\Gamma,\omega)\Rightarrow(\alpha_1,\Gamma_1,\omega_1)$ *and* $(\alpha,\Gamma,\omega)\Rightarrow(\alpha_2,\Gamma_2,\omega_2)$ *, then there exists a birooted word graph* $(\alpha',\Gamma',\omega')$ *such that* $(\alpha_1,\Gamma_1,\omega_1)\Rightarrow^* (\alpha',\Gamma',\omega')$ *and* $(\alpha_2,\Gamma_2,\omega_2)\Rightarrow^* (\alpha',\Gamma',\omega')$.

A birooted word graph is *closed* with respect to P if no P-production applies to *A*. One may use the Confluence Lemma to prove the following fact.

THEOREM 3. *Every birooted word graph A has a unique closure* $\overline{A}$ *with respect to every inverse presentation.*

We remark that $\overline{A}$ may be infinite even if *A* is finite. Let P be an inverse monoid presentation that will remain fixed in the following discussion. Let $u=x_1 \cdots x_n \epsilon (X\cup X^{-1})^*$. Define $\Gamma(u)$ to be the birooted word graph:

and let $B\Gamma_P(u)=\overline{\Gamma(u)}$, be the closure of $\Gamma(u)$ with respect to P. We will write $B\Gamma(u)$ if there is no ambiguity. Recall that an inverse monoid M has a natural partial order $\leq$ defined by: $m \leq n$ if and only if there is an idempotent e such that m=en.

THEOREM 4. (Stephen [18]) *Let* $M=Inv<X|u_i = v_i, i\epsilon I>= FIM(X)/\tau$. *Then for each* $u\epsilon(X\cup X^{-1})^*$, $L(B\Gamma(u))=\{w|w\tau\geq u\tau$ *in the natural partial order* $\geq$ on $M\}$.

As a consequence we obtain the following:

COROLLARY 5. *For* $u,v\epsilon(X\cup X^{-1})^*$ *the following are equivalent:*

(a) $u\tau=v\tau$ (*i.e.* $u=v$ *in M*).

(b) $u\epsilon L(B\Gamma(v))$ *and* $v\epsilon L(B\Gamma(u))$.

(c) $L(B\Gamma(u))=L(B\Gamma(v))$.

(d) $B\Gamma(u)=B\Gamma(v)$.

REMARK: There is a natural product on birooted word graphs. If $A=(\alpha,\Gamma,\omega)$ and $B=(\alpha',\Gamma',\omega')$ are birooted word graphs over X, then the product *AB* is obtained by concatenating *B* onto *A* by identifying the vertices ω and α'. The initial vertex of *AB* is α and the terminal vertex is ω'. Using this product, the product * in M is given by: $B\Gamma(u)*B\Gamma(v)=\overline{B\Gamma(u)B\Gamma(v)}$.

It is clear from Corollary 5 that the word problem for M is decidable if and only if there is an algorithm for deciding, for each $u,v\epsilon(X\cup X^{-1})^*$, whether $B\Gamma(u)=B\Gamma(v)$. Before providing some general sufficient conditions for this we give some examples that illustrate the construction of these graphs.

EXAMPLE 1. Consider the presentation $Inv<X|\emptyset>$ of the free inverse monoid on X. Since there are no relations, there are no expansions relative to this presentation. It is then easy to see that for each $u\epsilon(X\cup X^{-1})^*$, $B\Gamma(u)$ is a birooted *tree* and in fact is the Munn tree of u. (See [14], chapter VIII). Corollary 5 then specializes to:

THEOREM 6 (Munn[12]) *Elements of* FIM(X) *are in one to one correspondence with birooted deterministic word trees over* X.

EXAMPLE 2 Consider the presentation M=Inv<{a}|aa^{-1}=1> of the bicyclic monoid. It is easy to see that BΓ(1) is the birooted word tree shown in diagram 1.

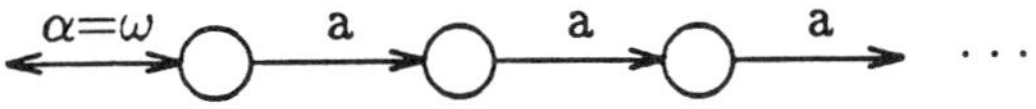

DIAGRAM 1

All birooted word graphs of the form BΓ(u) for $u \in \{a,a^{-1}\}^*$ may be obtained in a similar fashion (essentially by moving the start and terminal vertex to any other (possibly different) vertices). The usual solution to the word problem is obtained from Corollary 5.

EXAMPLE 3 Consider the presentation M=Inv<{a,b,c}| ab=ba,ac=ca,bc=cb> and let $u=b^{-1}bab^{-1}ca^{-1}bc^{-1}$. The Munn tree of Γ(u) is shown in diagram 2.

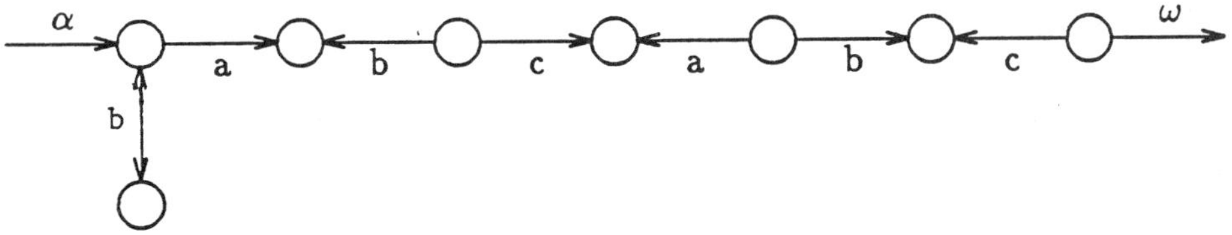

DIAGRAM 2

After performing all possible P-productions (in any order!) one checks that the birooted word graph BΓ(u) is given by diagram 3.

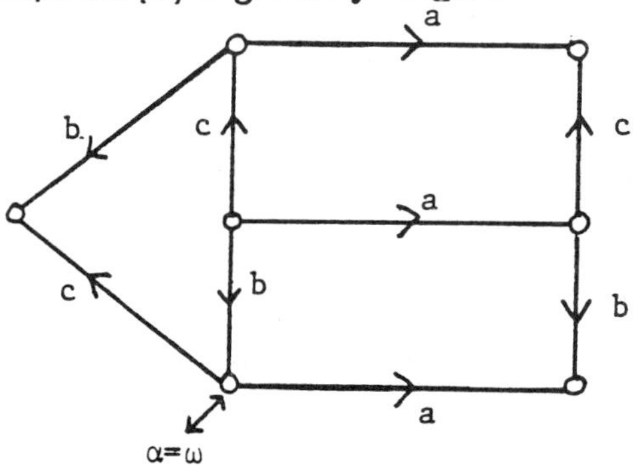

DIAGRAM 3

One can see from this for example, by constructing $B\Gamma(v)$ for the word $v=cb^{-1}c^{-1}aca^{-1}c^{-1}bab^{-1}a^{-1}b$, that $u\tau=v\tau$; that is u=v in M.

The Green's relations and other basic structural information about M may be obtained from the graphs $B\Gamma(u)$. Recall that E(M) denotes the semilattice of idempotents of M. We say that two paths in a graph are coterminal if their initial and terminal vertices are the same.

LEMMA 7. (Stephen [18]) *Let* $M=\mathrm{Inv}\langle X|R\rangle$ *and let* $G=\mathrm{Gp}\langle X|R\rangle$. *If* $u,v\in(X\cup X^{-1})^*$ *with* $B\Gamma(u)=(\alpha,\Gamma,\omega)$ *and* $B\Gamma(v)=(\alpha',\Gamma',\omega')$ *, then:*

(a) $u\tau\in E(M)$ *if and only if* $\alpha=\omega$.

(b) $u\tau R v\tau$ *if and only if there is a graph isomorphism* θ *from* Γ *onto* Γ' *such that* $\alpha\theta=\alpha'$.

(c) $u\tau L v\tau$ *if and only if there is a graph isomorphism* θ *from* Γ *onto* Γ' *such that* $\omega\theta=\omega'$.

(d) *If* π_1 and π_2 *are coterminal paths in* $B\Gamma(u)$ *labelling the words* u_1 and u_2 *respectively, then* $u_1=u_2$ *in G. That is,* $(u_1\tau)\sigma(u_2\tau)$ *where* σ *is the minimum group congruence on M.*

(e) *There is a one to one correspondence between the vertices of* $B\Gamma(u)$ *and the R-class of* $u\tau$ *in M.*

(f) *The transformation monoid of* $B\Gamma(u)$ *is the Schützenberger representation of M relative to the R class of* $u\tau$.

In addition, various conditions may be given that enable us to decide when $B\Gamma(u)=B\Gamma(v)$ for $u,v\in(X\cup X^{-1})^*$. We mention one such condition here. More general conditions are obtained in [18].

THEOREM 8. (Stephen [18]) *Let* $M=\mathrm{Inv}\langle X|u_i=v_i, 1\leq i\leq n\rangle$. *If* $u_i,v_i\in X^*$ and $\mathrm{length}(u_i)=\mathrm{length}(v_i)$ *for* $1\leq i\leq n$, *then* $B\Gamma(u)$ *is finite for all* $u\in(X\cup X^{-1})^*$. *In particular the word problem is solvable for this presentation.*

COROLLARY 9. *Let* $Mc_n=\mathrm{Inv}\langle\{x_1,\ldots,x_n\}\,|\,x_ix_j=x_jx_i, 1\leq i<j\leq n\rangle$. *Then the word problem for* Mc_n *is decidable.*

We remark that McAlister and McFadden [11] established this result for n=2. Their proof depended on the fact that Mc_2 is E-unitary but does not carry over to the general case since Mc_n is not E-unitary for n>2.

3. THE E-UNITARY PROBLEM: ONE RELATOR PRESENTATIONS

In this section we consider the E-unitary decision problem: when is M=Inv<X|R> E-unitary? As before we let M=Inv<X|$u_i = v_i$, $i \in I$>. Thus $G \simeq FG(X)/N$ where N is the normal subgroup of FG(X) generated by $\{u_i v_i^{-1}, i \in I\}$. By Lemma 1, G is the maximal group image of M (and FG(X) is the maximal group image of FIM(X)). Let ϕ denote the natural map from FIM(X) onto M. The proof of the following is routine.

LEMMA 10. *M is E-unitary if and only if, for each word* $u \in N$ *(regarded as a word in* FIM(X)*),* $u\phi$ *is an idempotent of M.*

For the remainder of the paper we restrict attention to the one relator case, that is M=Inv<X|w=1> for some $w \in (X \cup X)^{-1}$. The easiest case to consider is when w is an idempotent of FIM(X). For example, if X={a} and $w=aa^{-1}$, then M is the bicyclic monoid on M. The first part of the following theorem is immediate from Lemma 10.

THEOREM 11. (Margolis and Meakin [9], see also [8]). *Let* M=Inv<X|e=1> *where e is an idempotent of the free inverse monoid. Then*

(a) *M is E-unitary.*

(b) $B\Gamma(u)$ *is a birooted tree for all* $u \in (X \cup X^{-1})^*$ *and embeds in the tree of* FG(X). *(See Serre [17] for a discussion of trees and free groups).*

(c) $L(B\Gamma(u))$ *is a context free language for all* $u \in (X \cup X^{-1})^*$.

(d) *The word problem is decidable.*

Let us now return to the case M=Inv<X|w=1> where $w \in (X \cup X^{-1})^*$. We may think of w as an element of FIM(X). It is well known (e.g. [16]) that in FIM(X), we may write $w = e\rho(w)$ where $e \in E(FIM(X))$ and $\rho(w)$ is the reduced form of w (in the group theoretic sense). Now $e\rho(w)=1$ in M if and only if e=1 in M and $\rho(w)=1$, so if T=Inv<X|$\rho(w)=1$> is E-unitary, then M is an idempotent-pure image of T and hence M is E-unitary.

Partly for this reason, we restrict our attention to the case when w is a reduced word in the sense of group theory. That is w contains no segment of the form xx^{-1} for $x \in X \cup X^{-1}$. We have the following conjecture.

CONJECTURE 1. If w is a reduced word, then M is E-unitary if and only if w is *cyclically reduced.* That is, the first letter of w is not the inverse of the last letter of w.

As partial evidence in support of this conjecture we prove the following result.

THEOREM 12. *If* $w=uvu^{-1}$ where $u,v\epsilon X^+$ *and* u *and* v *end in different letters (so that* uvu^{-1} *is reduced as written), then M is not E-unitary.*

PROOF Consider the monoid N=Mon<X|uv=u>. By Adjan [1], N is a right cancellative monoid. On the other hand, $v\neq 1$ in N since the empty word is the only word in its congruence class and $v\epsilon X^+$. Thus v is not an idempotent of N.

It is well known that every right cancellative monoid K without idempotents except for 1 embeds as the R-class of 1 into its inverse hull $\Sigma(K)$. (See [2], Theorem 1.21). In particular, N embeds into $\Sigma(N)$ as the R-class of 1. Furthermore $\Sigma(N)$ is generated as an inverse monoid by X and satisfies the relation $uvu^{-1}=1$, since u and v are in the R-class of 1. Therefore, $\Sigma(N)$ is a quotient of M. This implies that v is not an idempotent of M since v is not an idempotent of $\Sigma(N)$. It follows that M is not E-unitary since v is the identity in the maximal group image of M.

This provides evidence in support of conjecture 1 in one direction. We now provide evidence in support of the converse statement. Assume from now on that w is a cyclically reduced word in $(X\cup X^{-1})^*$. In order to use Lemma 10 to prove that M is E-unitary, we need first to be able to decide when a reduced word $u\epsilon FG(X)$ is in N (and this is decidable by the classical result of Magnus [6]) and then decide whether every such word is an idempotent of M. We reformulate this problem in terms of Lyndon diagrams. We recall from combinatorial group theory that a *Lyndon diagram* for a cyclically reduced word w is an oriented planar map D with edges labelled by elements of $X\cup X^{-1}$ satisfying:

(L1) D is connected and simply connected and the product of the labels on the edges of a boundary cycle of D (in order) at some exterior vetex O is reduced without cancellation.

(L2) Every face (that is, region) of D is labelled by a cyclic conjugate of $w^{\pm 1}$. That is it is labelled by $w^{\pm 1}$ if we start at the appropriate vertex.

We refer the reader to Lyndon and Schupp [5] for a more precise definition of Lyndon diagrams for an arbitrary presentation and for the role that they play in combinatorial group theory. We recall that every reduced word in N labels the boundary cycle of some Lyndon diagram and that every boundary cycle of every Lyndon diagram of w labels a word in N. In other words, Lyndon diagrams provide a convenient geometric tool for studying membership in N. Thus from Lemma 10, in order to prove that M=Inv<X|w=1> is E-unitary (for w a cyclically reduced word) it suffices to show that every boundary cycle of every Lyndon diagram of w labels an idempotent of M. In order to state a basic result along these lines we introduce the following definition.

Let w be a cyclically reduced word over $X\cup X^{-1}$ and let M=Inv<X|w=1>. If w is factored in the form w=uv with $u,v\epsilon(X\cup X^{-1})^*$, then the word vu is a *cyclic conjugate* of w. It is easy to check that every cyclic conjugate of w is an idempotent in M. We say that a cyclic conjugate vu of w is *good* if vu=1 in M.

For example if X={a,b} and w=aba, then M is in fact a group so that all cyclic conjugates of w are good. On the other hand if w=ab, then M is the bicyclic monoid so that the cyclic conjugate ba is not good. Notice that if vu is a good cyclic conjugate of w, then u and v are both in the group of units of M.

THEOREM 13. (Margolis and Meakin [10]) *Let w be a cyclically reduced word and* M=Inv<X|w=1>. *Corresponding to each Lyndon diagram D of* w *and each edge e on $\partial(D)$ (the boundary of D) there is at most one face $F=F(e)$ with $e\epsilon\partial(F)$. The monoid M is E-unitary if, for every Lyndon diagram D of* w, *there is at least one edge $e\epsilon\partial(D)$ such that the cyclic conjugate of* $w^{\pm 1}$ *obtained by reading the labels of $\partial(F(e))$ in order starting with the edge e is a good cyclic conjugate of* $w^{\pm 1}$. *(We say that this good cyclic conjugate starts on the boundary of D).*

We illustrate many of the ideas developed so far in the following example.

EXAMPLE 4 Let X={a,b,c,d} and let w=abcdacdadabbcdacd. Consider the Lyndon diagram pictured in Diagram 4. The diagram has four faces F_1, F_2, F_3, F_4. The arrows at the vertices labelled 1,2,3,4 indicate the vertices on each face at which we start to read the word w. Thus if we start at vertex 1 [respectively 2,3,4] and read the label on $\partial(F_1)$[respectively $\partial(F_2)$,$\partial(F_3)$,$\partial(F_4)$] in a counterclockwise [respectively clockwise, counterclockwise, clockwise] direction we read the word w. Notice that on each face of this diagram the edge that is labelled by the first letter of w is not on ∂(D). At first sight this seems to suggest that the hypotheses of Lemma 13 are not satisfied. That is, the word obtained by reading around a boundary cycle of D may not be an idempotent of M and thus M may not be E-unitary.

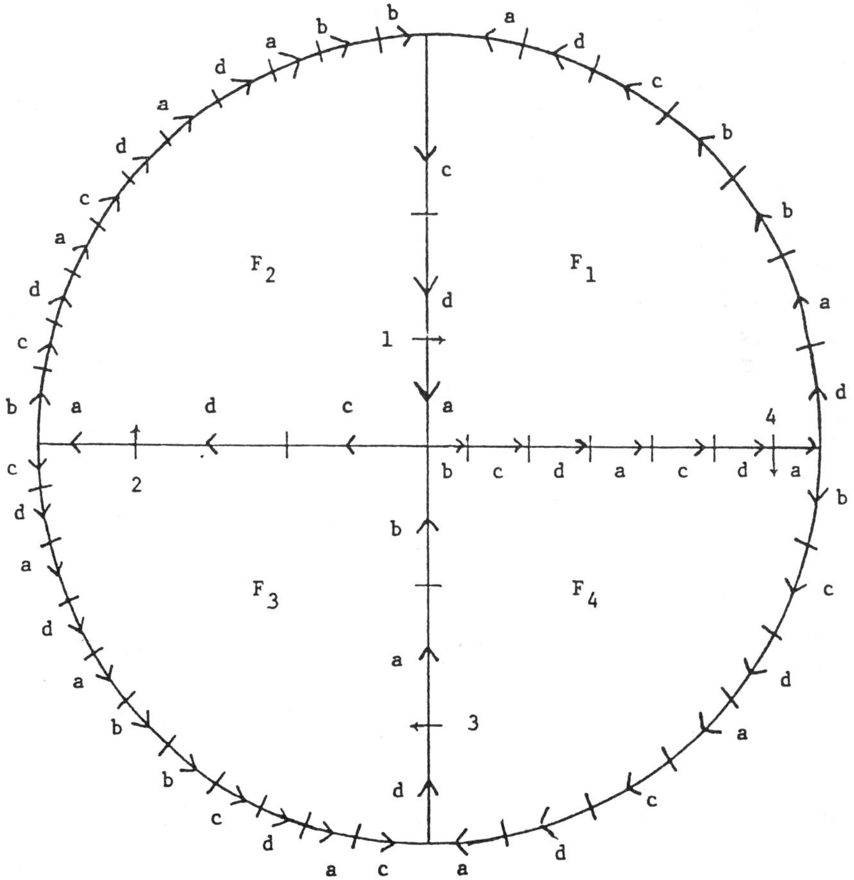

DIAGRAM 4

Notice however that the whole diagram is a finite approximate of $B\Gamma(1)$ starting at any of the vertices 1,2,3,4. We obtain the diagram D by starting at any one of these vertices and performing four P-expansions and reductions relative to this presentation. For example, we can start with the trivial graph at vertex 1, sew on the loop w so as to obtain the face F_1, then sew the loop w on at the vertex 4 and reduce so as to obtain the right hand half of D, then do the same thing at vertices 3 and 2 eventually obtaining the diagram D. It follows from the results of section 2 above, that *any* loop in this graph starting at 1 (or 2 or 3 or 4) must be labelled by a word whose value in M is 1. In particular, the word obtained by starting at 1 and reading around $\partial(F_2)$ in the clockwise direction must be 1 in M. This shows that the cyclic conjugate acdabcdacdadabbcd of w is a good cyclic conjugate of w. One sees by analyzing the loops at 1,2,3,4 that *all* the cyclic conjugates of w that start with the letter a are good. This shows that this particular Lyndon diagram has good cyclic conjugates that start on the boundary even though w itself does not start on the boundary of D. It shows more, since by the Freiheitsatz ([6] or [5], Theorem 5.1) every Lyndon diagram for w has at least one edge labelled by a on its boundary. Therefore some good cyclic conjugate of w starts on the boundary of every Lyndon diagram of w. By Lemma 13, we see that $M=Inv<\{a,b,c,d\}|abcdacdadabbcdacd=1>$ is E-unitary.

In particular the long word obtained by reading the boundary of D starting and ending at any vertex is an idempotent of M! It also follows that the group of units of M is non-trivial as there are many good cyclic conjugates of w. This is another remarkable feature about this example and points out the difference between inverse monoid presentations and monoid presentations. The monoid $N=Mon<\{a,b,c,d\}|abcdacdadabbcdacd=1>$ has trivial group of units: this follows from the results of Adjan [1]. This example also shows that while the search for members of the group of units is a "linear" problem in the case of monoid presentations (it is solved by analyzing overlaps between prefixes and suffixes of certain cyclic conjugates of w), the corresponding problem for inverse monoids is "two dimensional" in nature, depending on the geometry of Lyndon diagrams as in the example above.

We are now in a position to provide more evidence in support of our conjecture. First note the following Lemma which is easy to prove.

LEMMA 14 *Let* $M=Inv<X|w=1>$ *where w is a cyclically reduced word. Then M has trivial group of units if and only if w is the only good cyclic conjugate of itself.*

Using this and arguments involving Lyndon diagrams and the graph $B\Gamma(1)$ similar to the arguments presented in Example 4 we can prove the following.

THEOREM 15 (Margolis and Meakin [10]) *Let* $M=Inv<X|w=1>$ *where w is a cyclically reduced word. If M has trivial group of units, then M is E-unitary.*

We close the paper by discussing a class of examples of words w for which the corresponding monoid has trivial group of units and decidable word problem.

For each letter $x\epsilon X$ and each word $u\epsilon(X\cup X^{-1})^*$ we let $n_x(u)$ denote the sum of the exponents of x in u. Thus for example, $n_x(x^2)=2$, $n_x(x^{-1})=-1$ and $n_x(x^{-1}yx)=0$. We note that the map $n_x:(X\cup X^{-1})^*\to(Z,+)$ defined by $u\to n_x(u)$ for $u\epsilon(X\cup X^{-1})^*$ is a unary morphism. Also, if $u\rho v$ (where ρ is the Vagner congruence), then $n_x(u) = n_x(v)$ so that n_x may be considered as a morphism from FIM(X) to (Z,+).

Let X be a set with $|X|\geq 2$ and let w be a cyclically reduced word over X that involves at least two letters of $X\cup X^{-1}$. Let a be the first letter of w and let b be a letter of w distinct from a. Let $p=n_a(w)$ and $q=n_b(w)$. For each word $u\epsilon(X\cup X^{-1})^*$, let $k_b(u)=qn_a(u)-pn_b(u)$. Note that $k_b(w)$ is a morphism from FIM(X) to (Z,+). We say that the word w is *k-positive* if $k_b(u)>0$ for all prefixes u of w distinct from 1 and w (and some choice of b distinct from a). For example, it is easy to see that the word $w=aba^2b^3$ is k-positive but that the word $w=a^2b^3ab$ is not k-positive. We close the paper with the following result.

THEOREM 16 (Margolis and Meakin [10]) *Let w be a cyclically reduced word that is k-positive and let* M=Inv<X|w=1>. *Then:*

(a) *The group of units of M is trivial.*

(b) *M is E-unitary.*

(c) *The word problem for M is decidable.*

This result provides partial evidence for the following conjecture.

CONJECTURE 2 If M=Inv<X|w=1>, then the word problem for M is decidable.

REFERENCES

[1] S.I. Adjan,'Defining relations and algorithmic problems for groups and semigroups', Tr. Mat. Inst. Steklov **85** (Russian);Am. Math. Soc. (translation)**152**, 1967.

[2] A.H. Clifford and G.B. Preston, *The Algebraic Theory of Semigroups* ,Amer. Math. Soc., Math Surveys 7, Vol. 1,1961.

[3] S. Eilenberg, *Automata,Languages and Machines*,Vol.A,Academic Press, New York,1974.

[4] J.E. Hopcroft and J.D. Ullman, *Introduction to Automata Theory,Languages and Computation*,Addison-Wesley,1979.

[5] R.C. Lyndon and P.E. Schupp, *Combinatorial Group Theory*, Springer-Verlag, 1977.

[6] W. Magnus,'Das Identitäts-Problem für Gruppen mit einer definierenden Relation', Math. Ann.**106**,295-307,1932.

[7] S.W. Margolis and J.C. Meakin,'E-unitary inverse monoids and the Cayley graph of a group presentation', submitted for publication.

[8] S.W. Margolis and J.C. Meakin,'On a class of one relator inverse monoid presentations', Proc. L.S.U. Semigroup Conference, Louisiana State University, 1986, to appear.

[9] S.W. Margolis and J.C. Meakin,'Inverse monoids presented by idempotent relators', Manuscript in preparation.

[10] S.W. Margolis and J.C. Meakin,'On the E-unitary problem for inverse monoid presentations', Manuscript in preparation.

[11] D.B. McAlister and R. McFadden,'The free inverse semigroup on two commuting generators', J.Algebra,**32**,215-233,1974.

[12] W.D. Munn,'Free inverse semigroups', Proc. London Math. Soc. (3)**29**, 385-404,1974.

[13] M. Nivat and J.F.Perrot,'Une généralisation du monoïde bicyclique',C.R. Acad. Sci. Paris,**217A**,824-827,1970.

[14] M. Petrich, *Inverse Semigroups*, Wiley, 1984.

[15] C. Reutenauer,'Une topologie du monoïde libre', Semigroup Forum **18** ,No. 1,33-50,1979.

[16] H.E. Scheiblich,'Free inverse semigroups', Proc. Amer. Math. Soc. **38** 1-7,1973.

[17] J.-P. Serre, *Trees*,Springer-Verlag,1980.

[18] J.B. Stephen,'Presentations of inverse monoids', submitted for publication.

A CLASS OF INVERSE SEMIGROUP ALGEBRAS

W.D. Munn
Department of Mathematics
University of Glasgow
Glasgow G12 8QW
Scotland, U.K.

ABSTRACT. In 1976, Domanov showed that the algebra of an inverse semigroup S over a field F is semiprimitive (that is, has zero Jacobson radical) if the algebra of each maximal subgroup of S over F is semiprimitive. It is known that the converse statement is false in general. The principal purpose of this paper is to announce that if the semilattice of S satisfies a certain finiteness condition, introduced by Teply, Turman and Quesada in 1980, then the converse does hold. Corresponding results for primitivity are also discussed.

1. INTRODUCTION, NOTATION AND RING-THEORETIC PRELIMINARIES

We shall be concerned with the semigroup algebra of an inverse semigroup over a field. The study of semigroup algebras is largely motivated by that of group algebras and much of the interest derives from the interplay of two different disciplines: ring theory and semigroup theory. Since inverse semigroups have many group-like properties, one might expect to find results on inverse semigroup algebras not unlike those for group algebras. The main inspiration for the material presented here comes from a remarkable theorem by the Russian mathematician O.I. Domanov [3], quoted in the Abstract and stated as Theorem 4 below. A partial converse of this theorem is offered in §9.

The terminology and notation for semigroups is that of Clifford and Preston [2].

Let F be a field and let A denote an algebra over F. The Jacobson radical of A (that is, the largest quasiregular ideal of A) is denoted by J(A); and A is said to be <u>semiprimitive</u> if and only if J(A) = 0. A right A-module V is termed (i) <u>irreducible</u> if and only if V ≠ 0 and its only submodules are V and 0, (ii) <u>faithful</u> if and only if, for all a ∈ A, Va = 0 implies a = 0. We say that A is <u>primitive</u> if and only if A has a faithful irreducible right A-module. It can be shown that A is semiprimitive if and only if there exists a 'faithful family' of irreducible right A-modules, that is, a family of irreducible right A-modules such that the intersection of the annihilators of all the modules of the family is 0. In particular, from this it is clear that if A is

S. M. Goberstein and P. M. Higgins (eds.), Semigroups and Their Applications, 111–119.

primitive then A is semiprimitive. We note also that if A is finite-dimensional then the concept of semiprimitivity coincides with the classical notion of semisimplicity. (A finite-dimensional algebra is semisimple if and only if it has no nonzero nilpotent ideals.)

The semigroup algebra of a semigroup S over a field F, formally defined in [2, §5.2], is denoted here by F[S]. We may consider F[S] as a vector space over F with S as a basis: every nonzero element is uniquely expressible in the form $\sum_{i=1}^{n} \alpha_i x_i$, for some $n \in \mathbb{N}$, some distinct elements $x_1, x_2, \ldots, x_n$ of S and some nonzero elements $\alpha_1, \alpha_2, \ldots, \alpha_n$ of F; and multiplication in F[S] is induced in the natural way by the given multiplication in S. Thus the notion of a semigroup algebra is a straightforward generalisation of that of a group algebra.

2. GROUP ALGEBRAS

To provide a background, we begin by stating two well-known results on group algebras. The first of these, generally attributed to Maschke (1899), concerns the finite-dimensional case and underlies the classical theory of group representations.

THEOREM 1 (Maschke). *Let G be a finite group and let F be a field. Then F[G] is semisimple if and only if F has characteristic 0 or a prime not dividing the order of G.*

Next, we have the basic result on semiprimitivity of (possibly infinite-dimensional) group algebras. For the characteristic 0 case this is due to Amitsur [1] and for the prime characteristic case to Passman [12].

THEOREM 2 (Amitsur; Passman). *Let G be a group and let F be a field that is not an algebraic extension of its prime subfield. Suppose, further, that G has no element of order p if F has prime characteristic p. Then F[G] is semiprimitive.*

The hypothesis that F is not an algebraic extension of its prime subfield is satisfied, of course, if F is uncountable. In particular, $\mathbb{C}[G]$ and $\mathbb{R}[G]$ are semiprimitive - as was first shown by Rickart [16] in 1950, using analytic methods.

It is a long-standing conjecture that $\mathbb{Q}[G]$ is semiprimitive. This is known to be true for wide classes of groups G; but the general question remains open.

3. FINITE-DIMENSIONAL INVERSE SEMIGROUP ALGEBRAS

In the mid 1950s, soon after inverse semigroups were first studied, an analogue of Maschke's theorem for finite-dimensional inverse semigroup algebras was obtained, independently, by Oganesyan [11], Ponizovskiĭ [14] and Munn [5, 6]. (In fact, the last two authors were concerned with the problem of finding sufficient conditions on a field F and an *arbitrary* finite semigroup S for F[S] to be semisimple.) Implicit in their work is the following key result:

THEOREM 3. *Let* S *be a finite inverse semigroup and let* F *be a field. Then* F[S] *is semisimple if and only if, for each maximal subgroup* G *of* S, F[G] *is semisimple.*

Note that, since all the maximal subgroups of S contained in a given $\mathcal{D}$-class are isomorphic, the phrase 'each maximal subgroup G of S' could be replaced by 'one maximal subgroup G from each $\mathcal{D}$-class of S'.

From Theorems 3 and 1 we deduce

COROLLARY 3A (Oganesyan; Ponizovskiĭ; Munn). *Let* S *be a finite inverse semigroup and let* F *be a field. Then* F[S] *is semisimple if and only if* F *has characteristic* 0 *or a prime not dividing the order of a subgroup of* S.

Corollary 3A was the starting-point for the theory, due to Ponizovskiĭ and the author, of matrix representations and characters of finite inverse semigroups.

4. DOMANOV'S THEOREM

The next major step in this line of development was taken by O.I. Domanov with the publication in 1976 of the following result [3, Theorem 1]:

THEOREM 4 (Domanov). *Let* S *be an inverse semigroup and let* F *be a field. If, for each maximal subgroup* G *of* S, F[G] *is semiprimitive then* F[S] *is semiprimitive.*

The proof involves constructing a faithful family of irreducible right F[S]-modules from faithful families of irreducible right F[G]-modules, where G runs through a set of maximal subgroups of S comprising one such subgroup from each $\mathcal{D}$-class of S.

Theorems 4 and 2 then yield the corollary below. (Strangely, Domanov dealt only with the characteristic 0 case.)

COROLLARY 4A (Domanov). *Let* S *be an inverse semigroup and let* F *be a field that is not an algebraic extension of its prime subfield. Suppose, further, that no subgroup of* S *has an element of order* p *if* F *has prime characteristic* p. *Then* F[S] *is semiprimitive.*

In particular, $\mathbb{C}$[S] and $\mathbb{R}$[S] are semiprimitive.

An alternative proof of Corollary 4A, based on an investigation of nil ideals of F[S], has recently been obtained by the author [9].

From Domanov's proof of Theorem 4 we immediately obtain

THEOREM 5. *Let* S *be a bisimple inverse semigroup and let* F *be a field.*

If, for some maximal subgroup G *of* S, $F[G]$ *is primitive then* $F[S]$ *is primitive.*

For example, the algebra of the bicyclic semigroup over an arbitrary field is primitive - as is well known.

5. THE EXAMPLES OF TEPLY, TURMAN AND QUESADA

By a *Clifford semigroup* we mean a semilattice of groups; that is, an inverse semigroup whose semilattice is central.

In the course of a study of rings that are 'supplementary semilattice sums of subrings' (otherwise: 'semilattice-graded rings') — and evidently unaware of Domanov's work in [3] — Teply, Turman and Quesada [19] in 1980 established Theorem 4 for the special case in which S is a Clifford semigroup. Furthermore, they provided examples of a Clifford semigroup S and a field F such that $F[S]$ is semiprimitive while, for some maximal subgroup H of S, $F[H]$ is not semiprimitive. Thus the converse of Theorem 4 is false.

These examples can be described as follows.

Example 1. Let F be a field of prime characteristic p and let G be a group, with a subgroup H, such that $J(F[G]) = 0$, while $J(F[H]) \neq 0$. (Such groups exist! For instance, we may take G to be the wreath product $(\mathbb{Z}_p,+)\wr(\mathbb{Z},+)$; then G has a (normal) subgroup H which is a direct product of countably many copies of $\mathbb{Z}_p$ and the group algebras $F[G]$, $F[H]$ have the properties claimed [13, Ch.7, 4.12].) Let E be the semilattice consisting of the initial segment $[0, \omega]$ of the ordinals under the usual order. Write

$$G_\alpha = \begin{cases} \{\alpha\} \times H & \text{if } \alpha = \omega \\ \{\alpha\} \times G & \text{if } \alpha < \omega \end{cases} \qquad (\alpha \in E).$$

Thus each G_α is a group under componentwise multiplication. Now let $S = \cup_{\alpha \in E} G_\alpha$ and define a multiplication on S by

$$(\alpha, x)(\beta, y) = (\alpha\beta, xy).$$

Then S is a Clifford semigroup, with semilattice isomorphic to E and maximal subgroups G_α $(\alpha \in E)$. Moreover, $J(F[S]) = 0$; but $J(F[G_\omega]) \neq 0$.

Example 2. Let F, G, H be as in Example 1 and let $E = [0, \omega]$, with a new partial ordering, namely that defined by

$$\begin{cases} 0 < \alpha < \omega & \text{if } \alpha \notin \{0, \omega\}, \\ \alpha \nleq \beta \text{ and } \beta \nleq \alpha & \text{if } \alpha, \beta \notin \{0, \omega\} \text{ and } \alpha \neq \beta. \end{cases}$$

Take S as in Example 1, with the same rule of multiplication. Then, once again, $J(F[S]) = 0$, while $J(F[G_\omega]) \neq 0$.

Further examples to illustrate the failure of the converse of Theorem 4 have been provided by Ponizovskiĭ [15] and the author [10]. In particular, for any prime p, there exists a bisimple inverse monoid S whose unit group G is an infinite direct product of cyclic groups of order p, while F[S] is primitive for all fields F ([10]).

6. PSEUDOFINITE SEMILATTICES

In this section we focus attention on a particular class of semilattices with a certain finiteness property first examined in [19].

Let E be a semilattice and let $e, f \in E$. We say that e *covers* f if and only if $e > f$ and there is no g in E such that $e > g > f$. For all $e \in E$, write

$$\mathrm{cov}(e) = \{f \in E: \ e \text{ covers } f\}.$$

Clearly such a set might be empty.

Definition. A semilattice E is *pseudofinite* if and only if the following two conditions hold:
(i) for all $e, f \in E$ with $e > f$ there exists $g \in \mathrm{cov}(e)$ such that $g \geq f$;
(ii) for all $e \in E$, $|\mathrm{cov}(e)| < \infty$.

Condition (i) ensures that $\mathrm{cov}(e) = \emptyset$ if and only if e is the zero element of E.

In [19], pseudofinite semilattices were termed *m.u.-semilattices* ('maximal under'). Examples include finite semilattices, inversely well-ordered chains, and semilattices of free inverse semigroups of finite rank (see [7]). The semilattices in the examples in §5 are *not* pseudofinite: in Example 1, $\mathrm{cov}(\omega) = \emptyset$ and in Example 2, $\mathrm{cov}(\omega)$ is infinite.

It was shown in [19] that the converse of Theorem 4 holds for Clifford semigroups with pseudofinite semilattices. Specifically we have

THEOREM 6 (Teply, Turman and Quesada). *Let S be a Clifford semigroup with a pseudofinite semilattice and let F be a field. If F[S] is semiprimitive then, for each maximal subgroup G of S, F[G] is semiprimitive.*

A natural question at this stage is the following. Does Theorem 6 hold for an arbitrary inverse semigroup S with a pseudofinite semilattice? An affirmative answer is provided in §9. In fact the algebras of inverse semigroups with pseudofinite semilattices turn out to be relatively tractable. The key to the analysis of such algebras is provided by a concept introduced by A.V. Rukolaĭne in 1978 for the finite-dimensional case.

7. RUKOLAĬNE IDEMPOTENTS

Let S be an inverse semigroup with a pseudofinite semilattice E and let F be a field. For each e in E we have that $0 \leq |\mathrm{cov}(e)| < \infty$ and so we can define $\sigma(e) \in F[S]$ by

$$\sigma(e) = \begin{cases} \prod_{g \in \mathrm{cov}(e)} (e - g) & \text{if } e \text{ is not the zero of } E, \\ e & \text{otherwise.} \end{cases}$$

This notion, for the case in which S is a <u>finite</u> inverse semigroup, is due to Rukolaĭne [17]. Note that if $|\mathrm{cov}(e)| = n > 0$ then

$$\sigma(e) = e - s_1 + s_2 - s_3 + \ldots + (-1)^n s_n,$$

where s_r denotes the sum of all products of r distinct elements of cov(e) (r = 1, 2, ..., n).

It is evident that the elements $\sigma(e)$ $(e \in E)$ are nonzero idempotents of F[S]. They are also pairwise-orthogonal, as we now show. Let e, f be distinct elements of E and assume, without loss of generality, that $e \nless f$ under the natural partial ordering of E. Suppose first that $e > f$. Then there exists $g \in \mathrm{cov}(e)$ such that $g \geq f$. Hence $(e - g)f = 0$ and, for all $h \in \mathrm{cov}(f)$, $(e - g)(f - h) = 0$. Thus $\sigma(e)\sigma(f) = 0$. On the other hand, suppose that $e \ngtr f$. Then there exist $g \in \mathrm{cov}(e)$ and $h \in \mathrm{cov}(f)$ such that $g \geq ef$ and $h \geq ef$. Hence $(e - g)(f - h) = 0$ and so again $\sigma(e)\sigma(f) = 0$.

The 'Rukolaĭne idempotents' $\sigma(e)$ $(e \in E)$ have many other useful properties. In the case where S is finite these elements play a crucial part in the construction of a new basis for F[S] which clearly exhibits the structure of this algebra modulo group algebras [17, 18].

8. THE EXTENSION LEMMA

This brief section is devoted to the statement of a technical lemma required for the proof of Theorem 7 below.

Let S be an inverse semigroup, with pseudofinite semilattice E; let G be a maximal subgroup of S; let e be the identity of G; let D be the $\mathcal{D}$-class of S containing G; and let E(D) denote $E \cap D$. For all $f \in E(D)$ let t_f be an element of D such that $t_f t_f^{-1} = e$ and $t_f^{-1} t_f = f$.

With this notation we have the

LEMMA. <u>Let</u> F <u>be a field and let</u> A <u>be an ideal of</u> F[G]. <u>Write</u>

$$M(A) = \sum_{f,g \in E(D)} \sigma(f) t_f^{-1} A t_g \sigma(g).$$

<u>Then</u> M(A) <u>is an ideal of</u> F[S] <u>and is isomorphic to the algebra</u> $A_{E(D)}$ <u>of all</u> $E(D) \times E(D)$ <u>matrices over</u> A <u>with at most finitely many nonzero entries</u>.

The algebra operations in $A_{E(D)}$ are just the usual matrix operations. It can be shown that M(A) does not depend on the choice of the elements t_f ($f \in E(D)$).

9. THE MAIN RESULTS

The main results can now be formulated. First, we have a partial converse of Theorem 4 which generalises Theorem 6.

THEOREM 7. *Let S be an inverse semigroup with a pseudofinite semilattice and let F be a field. If F[S] is semiprimitive then, for each maximal subgroup G of S, F[G] is semiprimitive.*

We sketch a proof. Suppose that J(F[S]) = 0. Let G be a maximal subgroup of S, let D be the $\mathcal{D}$-class of S containing G and let A = J(F[G]). By the extension lemma (§8), F[S] contains an ideal $M(A) \cong A_{E(D)}$, where E(D) is the set of idempotents in D. Now A = J(A) and so

$$A_{E(D)} = (J(A))_{E(D)} = J(A_{E(D)}).$$

Hence $A_{E(D)}$ is quasiregular and so M(A) is quasiregular. Thus $M(A) \subseteq J(F[S])$. Consequently, M(A) = 0, which shows that $A_{E(D)} = 0$. It follows that A = 0.

Next, we state a partial converse of Theorem 5. This can also be proved with the help of the extension lemma.

THEOREM 8. *Let S be an inverse semigroup with a pseudofinite semilattice and let F be a field. If F[S] is primitive then S is bisimple and, for each maximal subgroup G of S, F[G] is primitive.*

Finally, we discuss briefly the algebra of a free inverse semigroup. It is known (see, for example, [7]) that the maximal subgroups of the free inverse semigroup S on a (nonempty) set X are all trivial and that S has infinitely many $\mathcal{D}$-classes; moreover, it follows easily from the description of the idempotents of S given in [7] that the semilattice of S is pseudofinite if and only if S has finite rank (that is, X is finite). Thus, from Theorems 4 and 8 we deduce

THEOREM 9. *Let S be a free inverse semigroup and let F be a field. Then*
(i) F[S] is semiprimitive;
(ii) if S has finite rank then F[S] is not primitive.

Part (i) was first established in [8]. The result in (ii) is unexpected: for Formanek [4] has shown that if S is a free group of rank not less than 2 or a free semigroup of rank not less than 2 then F[S] *is* primitive. At present the author does not know whether F[S] can be primitive when the rank of S is infinite.

10. REFERENCES

1. S.A. Amitsur. 'On the semi-simplicity of group algebras.' Michigan Math. J. 6 (1959), 251-253.

2. A.H. Clifford and G.B. Preston. The algebraic theory of semigroups. Math. Surveys 7 (American Math. Soc., Providence R.I., 1961 (Vol. I) and 1967 (Vol. II)).

3. O.I. Domanov. 'On the semisimplicity and identities of inverse semigroup algebras.' Rings and modules. Matem. Issled., Vyp. 38 (1976), 123-137. [In Russian].

4. E. Formanek. 'Group rings of free products are primitive.' J. Algebra 26 (1973), 508-511.

5. W.D. Munn. 'On semigroup algebras.' Proc. Cambridge Philos. Soc. 51 (1955), 1-15.

6. W.D. Munn. 'Matrix representations of semigroups.' Proc. Cambridge Philos. Soc. 53 (1957), 5-12.

7. W.D. Munn. 'Free inverse semigroups.' Proc. London Math. Soc. (3) 29 (1974), 385-404.

8. W.D. Munn. 'Semiprimitivity of inverse semigroup algebras.' Proc. Roy. Soc. Edinburgh A 93 (1982), 83-98.

9. W.D. Munn. 'Nil ideals in inverse semigroup algebras.' Submitted to J. London Math. Soc.

10. W.D. Munn. 'Two examples of inverse semigroup algebras.' Submitted to Semigroup Forum.

11. V.A. Oganesyan. 'On the semisimplicity of a system algebra.' Akad. Nauk Armyan. SSR Dokl. 21 (1955), 145-147. [In Russian].

12. D.S. Passman. 'On the semisimplicity of twisted group algebras.' Proc. Amer. Math. Soc. 25 (1970), 161-166.

13. D.S. Passman. The algebraic structure of group rings (Wiley - Interscience, New York, 1977).

14. I.S. Ponizovskiĭ. 'On matrix representations of associative systems.' Mat. Sbornik 38 (1956), 241-260. [In Russian].

15. I.S. Ponizovskiĭ. 'An example of a semiprimitive semigroup algebra.' Semigroup Forum 26 (1983), 225-228.

16. C. Rickart. 'The uniqueness of norm problem in Banach algebras.' Ann. Math. 51 (1950), 615-628.

17. A.V. Rukolaĭne. 'The centre of the semigroup algebra of a finite inverse semigroup over the field of complex numbers.' Rings and linear groups. Zap. Naučn. Sem. Leningrad. Otdel. Mat. Inst. Steklov (LOMI) 75 (1978), 154-158. [In Russian].

18. A.V. Rukolaĭne. 'Semigroup algebras of finite inverse semigroups over arbitrary fields.' Modules and linear groups. Zap. Naučn. Sem. Leningrad. Otdel. Mat. Inst. Steklov (LOMI) 103 (1980), 117-123. [In Russian].

19. M.L. Teply, E.G. Turman and A. Quesada. 'On semisimple semigroup rings.' Proc. Amer. Math. Soc. 79 (1980), 157-163.

ACKNOWLEDGEMENT

I wish to thank the organisers of the Chico Conference on Semigroups, Dr. S.M. Goberstein and Dr. P.M. Higgins, for inviting me to take part and for their kind hospitality. I must also record my gratitude to California State University, Chico for providing me with generous financial support.

CATEGORICAL EXTENSION THEORY -- REVISITED

William R. Nico
California State University, Hayward
Department of Mathematics & Computer Science
25800 Carlos Bee Boulevard
Hayward, CA 94542

The study of extensions of semigroups with identity (monoids) leads naturally to the study of extensions of categories. The fundamental structure is developed from the study of group extensions in the context of wreath products done by Krasner and Kaloujnine [1].

This investigation was motivated by two fundamental questions:
1. How can one state an extension problem for monoids, i.e., what is a kernel?
2. Why have wreath products come to play such a pivotal role in the study of (finite) semigroups?

This is an attempt to show that the answers to the two questions are related. Most details of what follows can be found in [2].

One main idea is that since it appears to be reasonable to describe the kernel of a homomorphism of monoids as a category, it is also convenient to consider a monoid as a category of one object. In this context, what are ordinarily thought of as the elements of a monoid become the arrows of the category. Furthermore, one finds that there is no reason to restrict consideration to categories of only one object.

The second main idea comes from the work of Krasner and Kaloujnine, who proved in [1] that every group extension with quotient G and kernel N had a canonical embedding into the standard wreath product $G \text{ w } N$. Moreover, two group extensions which are equivalent in the sense of Schreier have embeddings which are conjugate to each other by an element of the form $(1, n)$ in the wreath product.

Since it is possible to build a wreath product of categories, it is possible to study extensions of categories (and hence of monoids) in terms of their canonical embeddings in such wreath products. It is also possible to describe a pre-order on such extensions with a relation of conjugation. However, since categories generally lack inverses to arrows (as do monoids), this pre-order will not separate the collection of extensions into equivalence classes, as happens in the case of groups.

S. M. Goberstein and P. M. Higgins (eds.), Semigroups and Their Applications, 121–124.

Call a morphism $\pi : A \twoheadrightarrow B$ of categories which is full and bijective on objects a quotient functor. Then the kernel of π can be described as a category K having as objects the arrows of B and having arrows $(a,b) : b \to b'$ where b and b' are arrows of B, a is an arrow of A and the diagram below commutes.

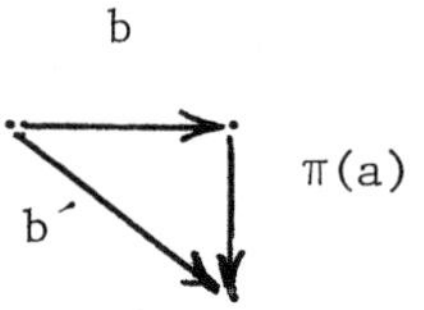

A similar structure, described in matrix terms and called the derived semigroup, was discovered independently by Tilson (see [3]).

The first result is that there is a natural embedding of the category A into the wreath product $B \, w \, K$ of categories.

To describe briefly the wreath product of categories, consider a category C and a functor $F : C^{op} \to$ Cat, where Cat is the category of small categories. The semidirect product $C \times F$ is the category with objects pairs (x, u), where $x \in |C|$, and $u \in |F(x)|$. Then an arrow $(c, p) : (x, u) \to (y, v)$ consists of an arrow $c : x \to y$ in C and an arrow $p : u \to F(c)(v)$ in the category $F(c)$. If one writes

q^c for $F(c)(q)$ when q is an arrow in $F(y)$, then

composition of arrows in $C \times F$ becomes $(d, q)\ (c, p) = (dc, q^c p)$.

To construct a standard wreath product ${}_R C \, w \, D$ of two categories C and D form the semidirect product $C \times D^R$,

where $D^R : C \to$ Cat is built from the right regular representation R of C on the category of sets. (For an object x of C, $R(x)$ is the set of all arrows of C with codomain x. If $R(x)$ is thought of as a discrete category, then $D^{R(x)}$ is the category of all functors from $R(x)$ to D.)

The problem which arises is that the semidirect product $C \times F$, and hence the wreath product $C \, w \, D$, have too many objects for the natural functors which project them onto the category C, e.g., $C \times F \twoheadrightarrow C$, to be bijective on objects as required of a quotient functor. The solution is to have a choice function ϕ from objects of C to objects of $C \times F$ (or $C \, w \, D$) and then use only the full subcategory generated by the chosen objects. Doing this for an arbitrary choice function can yield a full subcategory (e.g., $C \times_\phi F$) such that the natural restriction of the projection is indeed bijective on objects, but fails to be full, still not fulfilling the requirements of a choice function. Since a choice function here is simply a functor from the discrete category $|C|$ to $C \times F$ (or to $C \, w \, D$), we may refer to a choice functor ϕ such that the natural projection $C \times_\phi F \to C$ (or $C \, w_\phi \, D \to C$) is a quotient functor as a compatible functor.

It is worth noting here that a choice functor such as is needed in the case of a wreath product $C \, w \, D$ is, after all the technicalities are stripped away, simply a function from the set of arrows of C, arr C, to the objects of D, which we may write as a functor $\phi : \text{arr } C \dashrightarrow D$. Then ϕ is a compatible functor if, for every pair of composable arrows s, t in C, one has $D(\phi(t), \phi(st))$ non-empty.

Given categories C and D and a compatible functor ϕ, as above, then an extension of C by (ϕ, D) is a category A together with a quotient morphism $\pi : A \dashrightarrow C$ and an embedding $\delta : A \dashrightarrow C \, {}_{\phi}w \, D$ such that the diagram

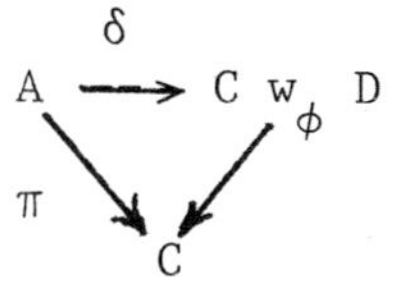

A more explicit description of this embedding can be had by observing that the functor δ must be of the form $\delta = (\pi, \phi)$ on objects and of the form $\delta = (\pi, \lambda)$ on arrows, where for each arrow a of A, $\lambda(a)$ is a natural transformation $\phi \dashrightarrow \phi^{\pi(a)}$. That is, for every arrow c of C for which the composition $\pi(a)c$ is defined, $\lambda(a)_c$ is an arrow $\lambda(a)_c : \phi(c) \dashrightarrow \phi(\pi(a)c)$ in D. This λ must satisfy two conditions:

1. $\lambda(aa')_c = \lambda(a)_{\pi(a')c} \lambda(a)_c$ and

2. λ and π must be "jointly one-to-one" in a sense that makes δ an embedding.

The main reason for thinking in terms of this λ is the similarity of condition 1, to a cocycle condition. For this reason, the author refers to λ as the cocycle of the extension.

One can say that one extension (A_1, π_1, δ_1) is conjugate to another extension (A_2, π_2, δ_2) if there is a morphism $\sigma : A_1 \dashrightarrow A_2$ and a natural transformation $\rho : \delta_1 \dashrightarrow \delta_2\sigma$ as in the diagram below.

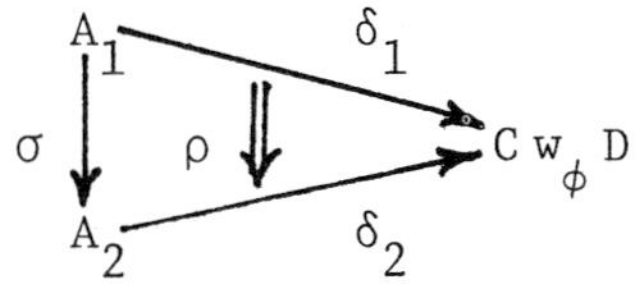

If one looks at the decomposition of δ into the form $\delta = (\pi, \lambda)$ on arrows, it becomes clear that ρ can be decomposed into the form $\rho = (1, \mu)$, where $\mu : \phi \dashrightarrow \phi$ is a natural transformation making the square below commute.

$$\begin{array}{ccc} \phi & \xrightarrow{\lambda(a)} & \phi^{\pi(a)} \\ \mu \downarrow & & \downarrow \mu^{\pi(a)} \\ \phi & \longrightarrow & \phi^{\pi(a)} \end{array}$$

It is this diagram which accounts for calling the relationship *conjugacy*.

This viewpoint provides a rich context in which to investigate extensions of monoids and relationships between such extensions. It also leads the author to speculate that complexity theory for finite semigroups as developed by J. Rhodes, et al., (see [3]) could be developed in terms of extension and conjugacy described briefly here. Specifically the author conjectures that the fundamental relation *divides*, well known for semigroups and extended by Tilson to categories, can be given a precise interpretation to the effect that $A \mid B$ means that both are extensions of some C by some (ϕ, D) and that A is conjugate to B (or perhaps vice versa).

References

1. M. Krasner and L. Kaloujnine, 'Produit complet des groupes de permutations et problème d'extension de groupes,' I, II, III, *Acta. Sci. Math. Szeged*, **13**(1950), 208-230; **14**(1951), 39-66; **14**(1951), 69-82.

2. W. Nico, 'Wreath Products and Extensions,' *Houston Journal of Mathematics*, **9**(1983), 71-99.

3. B. Tilson, 'On the Complexity of Finite Semigroups,' *J. Pure and Appl. Algebra*, **5**(1974), 187-208.

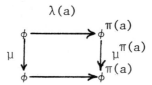

A CLASS OF UNIFORM CHAINS

Francis Pastijn
Department of Mathematics, Statistics and Computer Science
Marquette University
Milwaukee, WI 53233

ABSTRACT. A chain (that is, a linear ordering) is said to be uniform if all of its principal ideals are isomorphic. A chain is scattered if it does not contain a subchain which is isomorphic to the chain of the rational numbers. We shall construct a class of pairwise non-isomorphic scattered uniform chains.

1. INTRODUCTION

In the following we shall use the notation and terminology of [6]. The study of uniform chains is to some extent motivated by the investigations into the structure of bi-simple inverse semigroups (see e.g. [4]). A preliminary study of scattered uniform chains was begun in [5].

Let A be a scattered uniform chain. For each ordinal α we define a congruence relation ρ_α on A. The congruences ρ_α are defined inductively as follows :

(i) ρ_0 is the equality on A,

(ii) if β is an ordinal, then $a\,\rho_{\beta+1}\,b$ if and only if there are only finitely many elements of A/ρ_β between $a\rho_\beta$ and $b\rho_\beta$,

(iii) if α is a limit ordinal, then $\rho_\alpha = \bigcup_{\beta<\alpha} \rho_\beta$

There exists a least ordinal α such that A/ρ_α is the one-element chain. This ordinal α will be called the rank of A, in notation $\alpha = r(A)$ (called the F-rank in [6]).

The chain Z of integers and the chain N^* of natural numbers under the reverse ordering are, up to isomorphism, the only scattered uniform chains of rank 1. Their order types are denoted $\overline{Z} = \zeta$ and $\overline{N^*} = \omega^*$. A transitive chain is obviously uniform. A scattered uniform chain is transit-

S. M. Goberstein and P. M. Higgins (eds.), Semigroups and Their Applications, 125–132.

ive of rank α if and only if its order type is ζ^{α} (see [3] and 8.1 of [6]). A uniform dually well-ordered chain has order type $(\omega^{\alpha})^*$ for some ordinal α. Conversely, if α is any ordinal, then $(\omega^{\alpha})^*$ is the order type of a scattered uniform chain of rank α ([1],[2],[7]). In the following we shall give a structure theorem and a corresponding isomorphism theorem for all scattered uniform chains of finite rank.

Let A be a scattered uniform chain of rank α. For $a \in A$ we use the notation $a\rho_{\beta} = c_A^{\beta}(a)$. Thus $c_A^0(a) = \{a\}$ and α is the least ordinal such that $c_A^{\alpha}(a) = A$. If B is any subchain of A, and $a \in B$, then $c_B^{\beta}(a)$ does not necessarily coincide with $c_A^{\beta}(a) \cap B$. However, if $B = c_A^{\gamma}(a)$ for some $\beta \leq \gamma \leq \alpha$, then $c_B^{\beta}(a) = c_A^{\beta}(a)$ and the ρ_{β}-relation on B is the restriction to B of the ρ_{β}-relation on A. Since we shall deal with such settings only we can omit the subscripts A and B and we shall henceforth write $c^{\beta}(a)$ instead of $c_A^{\beta}(a)$ or $c_B^{\beta}(a)$. Also the double use of ρ_{β} in the above will not cause any ambiguity. Further, for $\gamma \leq \alpha$, $c^{\gamma}(a)$ is a scattered uniform chain of rank γ (Proposition 2.5 of [5]). Also if $a,b \in A$ and if $c^{\gamma}(a)$ and $c^{\gamma}(b)$ both have immediate successors in A/ρ_{γ}, then $c^{\gamma}(a) \cong c^{\gamma}(b)$ (Proposition 2.3 of [5]).

2. MAIN RESULTS

In the following an ordinal α will stand for the chain of ordinals which precede α. A transformation $f : \alpha \to \alpha$ is said to be <u>non-increasing</u> if $f(\beta) \leq \beta$ for every $\beta < \alpha$.

<u>Construction</u> 1. Let α be an ordinal and f a non-increasing transformation of $\alpha+1$ such that $f(\gamma)$ is not a limit ordinal for every $\gamma \leq \alpha$. We construct the chains A_{γ}, $\gamma \leq \alpha$, inductively as follows :

(i) $A_0 = 1$,

(ii) if $f(\gamma) = 0$ and $\gamma > 0$, then

$$A_{\gamma} = \Sigma \{A_{\beta}N^* \mid \beta \in \gamma^*\},$$

(iii) if $f(\gamma) = \xi > 0$, then

$$A_\gamma = \Sigma\ \{\ A_\beta N^* \mid \beta \in [\xi,\gamma)^*\ \} + A_{\xi-1}Z \quad .$$

The chain A_α will also be denoted by A_f. Obviously the restriction $f|_{\gamma+1}$ is a non-increasing transformation of $\gamma+1$ for every $\gamma \leq \alpha$, and $A_\gamma = A_{f|_{\gamma+1}}$. This chain has a greatest element if and only if $f(\gamma) = 0$.

If f is as above, then we define $\bar{f}$ to be the transformation of $\alpha+1$ such that $\bar{f}(\gamma) = f(\gamma)$ if $\gamma < \alpha$ and $f(\alpha) = 0$. Evidently $A_{\bar{f}}$ has a greatest element.

Proposition 2. Let α be an ordinal and f a non-increasing transformation of $\alpha+1$ whose values are not limit ordinals. Then every principal ideal of A_f is isomorphic to $A_{\bar{f}}$.

Proof. We show by induction that for every $\gamma \leq \alpha$ we have that the principal ideals of A_γ are isomorphic to $A_{\overline{f|_{\gamma+1}}}$. This is obviously the case if $\gamma = 0$.

We next assume that $\gamma > 0$ and that for $\beta < \gamma$ we have that the principal ideals of A_β are isomorphic to $A_{\overline{f|_{\beta+1}}}$. If $f(\gamma) = \xi \neq 0$, then

$$A_\gamma = \Sigma\ \{\ C_\beta \mid \beta \in [\xi-1,\gamma)^*\ \}$$

where

$$C_\beta = \begin{cases} A_\beta N^* & \text{if } \beta \in [\xi,\gamma)^* , \\ A_{\xi-1}Z & \text{if } \beta = \xi-1 . \end{cases}$$

We choose $a \in A_\gamma$ and calculate the order type of the principal ideal $A_\gamma^{\leq a}$. There exists $\theta \in [\xi-1,\gamma)^*$ such that $a \in C_\theta$, so that

$$A_\gamma^{\leq a} = \Sigma\ \{\ C_\beta \mid \beta \in [\theta+1,\gamma)^*\ \} + C_\theta^{\leq a} \quad .$$

By the induction hypothesis,

$$\begin{aligned} C_\theta^{\leq a} &\cong A_\theta N^* + A_{\overline{f|_{\theta+1}}} \\ &= A_\theta N^* + \Sigma\ \{\ A_\beta N^* \mid \beta \in \theta^*\ \} \quad , \end{aligned}$$

hence

$$A_\gamma^{\leq a} \cong \Sigma\ \{\ A_\beta N^* \mid \beta \in \gamma^*\ \} = A_{\overline{f|_{\gamma+1}}} \quad .$$

In particular, since $f|_{\alpha+1} = f$, we have that every principal ideal of A_f is isomorphic to $A_{\bar{f}}$.

Corollary 3. Let α and f be as in Proposition 2. Then

A_f <u>is a scattered uniform chain of rank</u> α .

<u>Proof</u>. From the foregoing Proposition it follows that for every $\gamma \leqslant \alpha$, every principal ideal of A_γ is isomorphic to $A_{\overline{f|_{\gamma+1}}}$, hence A_γ is uniform. In particular A_f is uniform. A routine inductive argument shows that for every $\gamma \leqslant \alpha$, A_γ is scattered and has rank γ .

<u>Proposition</u> 4. <u>Let</u> α <u>be an ordinal and</u> f <u>and</u> g <u>non-increasing transformations of</u> $\alpha+1$ <u>whose values are not limit ordinals</u>. <u>Then</u> $A_f \cong A_g$ <u>if and only if</u> $f = g$.

<u>Proof</u>. Using the transformation f , we construct the A_γ , $\gamma \leqslant \alpha$, as in Construction 1 . For every $\gamma < \alpha$ and $a \in A_f$, $c^{\gamma+1}(a)/\rho_\gamma$ is of order type ζ or ω^* (Lemma 2.4 of [5]). Therefore for every $\gamma < \alpha$ we can choose $a_\gamma \in A_f$ such that $c^\gamma(a_\gamma)$ has an immediate successor in A_f/ρ_γ . We choose any $a_\alpha \in A_f$.

We prove by induction on α that $c^\gamma(a_\gamma) \cong A_\gamma$ for every $\gamma \leqslant \alpha$. This is obviously true if $\alpha = 0$. We next assume that $\alpha > 0$ and consider $\gamma < \alpha$. Let $f(\alpha) = \xi > 0$. Then

$$A_\alpha = \Sigma\ \{\ C_\beta \mid \beta \in [\xi-1,\alpha)^*\ \}$$

where

$$C_\beta = \begin{cases} A_\beta N^* & \text{if } \beta \in [\xi,\alpha)^* , \\ A_{\xi-1} Z & \text{if } \beta = \xi-1 . \end{cases}$$

By Corollary 3, C_β is of rank $\beta+1$ for every $\beta \in [\xi-1,\alpha)^*$. Hence if $\gamma < \alpha$, then $a_\gamma \in C_\theta$ for some $\theta \geqslant \gamma$ because otherwise $c^\gamma(a_\gamma)$ would be the greatest element of A_f/ρ_γ . Thus a_γ belongs to a copy of $A_\theta N^*$ or $A_\theta Z$, where $\gamma \leqslant \theta < \alpha$, with $A_\theta = A_{f|_{\theta+1}}$. By the induction hypothesis we have that $c^\gamma(a_\gamma) \cong A_\gamma$. The case $f(\alpha) = 0$ can be treated in a similar way. If $\gamma = \alpha$, then of course $A_f = A\alpha = c^\alpha(a_\alpha)$.

We next show that for every $\gamma \leqslant \alpha$, $f(\gamma)$ is the least element of $\alpha+1$ such that $c^\gamma(a_\gamma)/\rho_{f(\gamma)}$ has a greatest element. If $c^\gamma(a_\gamma) \cong A_\gamma$ has a greatest element, then indeed $f(\gamma) = 0$. Otherwise $\gamma \geqslant f(\gamma) > 0$ and

$$c^\gamma(a_\gamma) \cong A_\gamma = \Sigma\{ A_\beta N^* \mid \beta \in [f(\gamma),\gamma)^* \} + A_{f(\gamma)-1}Z .$$

By Corollary 3, $A_{f(\gamma)-1}$ is uniform of rank $f(\gamma)-1$, so that $A_{f(\gamma)-1}Z$ is uniform of rank $f(\gamma)$ and $A_{f(\gamma)-1}Z/\rho_{f(\gamma)-1} \cong Z$. It follows that $c^\gamma(a_\gamma)/\rho_\xi$ does not have a greatest element if $\xi < f(\gamma)$, whereas $c^\gamma(a_\gamma)/\rho_{f(\gamma)}$ has a greatest element.

If $\psi : A_f \to A_g$ is an isomorphism, then we put $\psi(a_\gamma) = b_\gamma$ for every $\gamma \leq \alpha$. Evidently $c^\gamma(b_\gamma)$ has an immediate successor in A_g/ρ_γ for every $\gamma < \alpha$, and $c^\gamma(b_\gamma) \cong c^\gamma(a_\gamma)$. By the above reasoning we may conclude that for every $\gamma \leq \alpha$, $f(\gamma) = g(\gamma)$ is the least element of $\alpha+1$ such that $c^\gamma(a_\gamma)/\rho_{f(\gamma)} \cong c^\gamma(b_\gamma)/\rho_{f(\gamma)}$ has a greatest element. Thus $f = g$.

Theorem 5. If α is an infinite ordinal, then there exist exactly $2^{|\alpha|}$ pairwise non-isomorphic chains with greatest element which are scattered, uniform and of rank α.

Proof. There exist $2^{|\alpha|}$ non-increasing transformations of $\alpha+1$ whose values are not limit ordinals and which take the value 0 in α. With every such transformation f there corresponds a chain A_f with greatest element, through Construction 1. The theorem now follows from Corollary 3 and Proposition 4.

Example 6. Let α be any ordinal, and let f be the transformation of $\alpha+1$ which is given by $f(\gamma) = 0$ for every $\gamma \leq \alpha$. Then the chain $A_\alpha = A_f$ of Construction 1 has order type $(\omega^\alpha)^*$.

By no means is it possible to construct an isomorphic copy of every scattered uniform chain via Construction 1. This is for instance the case with a chain of order type ζ^ω, which is a scattered uniform chain of rank ω.

Theorem 7. Let n be a finite ordinal and let f be a non-increasing transformation of $n+1$. Then the chain A_f of Construction 1 is a uniform chain of rank n. Conversely, every uniform chain of rank n is isomorphic to a chain constructed like this. If g is any other non-increasing transformation of $n+1$ then $A_f \cong A_g$ if and only if $f = g$.

Proof. It suffices to give a proof of the converse part. We let A be any scattered uniform chain of rank n. We shall assume that $n \geq 1$, the case $n = 0$ being evident. For every $a \in A$, we have that $c^{i+1}(a)/\rho_i$, $0 \leq i < n$, is of order type ω^* or ζ (Lemma 2.4 of [5]). It is therefore possible to choose $a \in A$ in such a way that $c^i(a)$ has a successor in $c^{i+1}(a)/\rho_i$ for every $i < n$. For $i \in n+1$ we define $f(i)$ to be the least element of $n+1$ such that $c^i(a)/\rho_{f(i)}$ has a greatest element. Obviously $f(i) \leq i$ since $c^i(a)/\rho_i$ is the one-element chain.

We may use the transformation f of $n+1$ to construct the chains $A_0, A_1, \ldots, A_n$ as in Construction 1. We shall show by induction that $c^i(a) \cong A_i$ for every $i \in n+1$. This is obviously the case if $i = 0$. We next assume that $i > 0$ and that $c^k(a) \cong A_k$ whenever $k < i$. We can choose $b \in c^i(a)$ such that $c^{f(i)}(b)$ is the greatest element of $c^i(a)/\rho_{f(i)}$. By Proposition 2.3 and Lemma 2.4 of [5]

$$c^j(b) \cong c^{j-1}(a)N^* + c^{j-1}(b) \quad \text{if} \quad f(i) < j \leq i$$

and

$$c^{f(i)}(b) = c^{f(i)-1}(a)Z \quad \text{if} \quad f(i) > 0 \quad .$$

Hence, if $f(i) < j \leq i$, then

$$c^i(a) \cong c^i(b) \cong c^{i-1}(a)N^* + \ldots + c^{f(i)}(a)N^* + c^{f(i)}(b) \quad .$$

In particular, if $f(i) = 0$, then $c^{f(i)}(a)N^* + c^{f(i)}(b) \cong N^*$ and the induction hypothesis gives

$$c^i(a) \cong A_{i-1}N^* + \ldots + A_1N^* + A_0N^* = A_i$$

as required. If $f(i) > 0$, then we have from the above that

$$\begin{aligned} c^i(a) &= c^i(b) \\ &\cong c^{i-1}(a)N^* + \ldots + c^{f(i)}(a)N^* + c^{f(i)-1}(a)Z \\ &\cong A_{i-1}N^* + \ldots + A_{f(i)}N^* + A_{f(i)-1}Z = A_i \end{aligned}$$

as required. In particular $A = c^n(a) = A_n$.

The above theorem allows us to set up a one-to-one correspondence between the non-increasing transformations of n+1 and the order types of scattered uniform chains of rank n . Therefore we have

Corollary 8 (Corollary 3.4 of [5]). There exist precisely n+1 ! order types of scattered uniform chains of rank n .

Example 9. Let f be the non-increasing transformation of 5 which is given by

$$f(0) = 0 , f(1) = 1 , f(2) = 1 , f(3) = 2 , f(4) = 1 .$$

We calculate $A_f = A_4$ gradually as follows :

$A_0 = 1$,
$A_1 = A_0 Z$ has order type ζ ,
$A_2 = A_1 N^* + A_0 Z$ has order type $\zeta\omega^* + \zeta$,
$A_3 = A_2 N^* + A_1 Z$ has order type $(\zeta\omega^* + \zeta)\omega^* + \zeta^2$,
$A_4 = A_3 N^* + A_2 N^* + A_1 N^* + A_0 Z$ has order type

$$((\zeta\omega^* + \zeta)\omega^* + \zeta^2)\omega^* + (\zeta\omega^* + \zeta)\omega^* + \zeta\omega^* + \zeta .$$

The proof of the following theorem is now routine.

Theorem 10. Let f be a non-increasing transformation of n+1 . Then

(i) A_f is dually well-ordered if and only if $f(i) = 0$ for every $i \in n+1$, and then $\overline{A_f} = (\omega^n)^*$,

(ii) A_f has a transitive automorphism group if and only if $f(i) = i$ for every $i \in n+1$, and then $\overline{A_f} = \zeta^n$.

From Theorem 5 we know that there exist $2^{\aleph_0}$ pairwise non-isomorphic chains with greatest element which are scattered, uniform and of rank ω . We now show that the construction used yields all such chains.

Theorem 11. Let f be a non-increasing transformation of $\omega+1$ such that $f(\omega)=0$. Then A_f is a chain with greatest element which is scattered, uniform and of rank ω. Every such chain can be so constructed.

Proof. It suffices to give a proof of the last part of the theorem. So, let A be a scattered uniform chain of rank ω with a greatest element a. For every $i<\omega$, $c^i(a)$ is a scattered uniform chain of rank i, and a is the greatest element of $c^i(a)$. There exists a unique transformation f_i of $i+1$ such that $c^i(a) = A_{f_i}$. It is easy to show that for any $0<i<j<\omega$, $f_j|_j$ extends $f_i|_i$. We define f to be the non-increasing transformation of $\omega+1$ which extends $\bigcup_{0<i<\omega} f_i|_i$ and for which $f(\omega)=0$. Using f we construct the A_i, $i<\omega$, as in Construction 1. Then

$$c^i(a) = A_{f_i} = A_{i-1}N^* + \ldots + A_0N^*$$

for every $i<\omega$ and

$$A = \bigcup_{i<\omega} c^i(a) = A_f = \Sigma\{\, A_iN^* \mid i \in \omega^* \,\} .$$

Example 12. Every principal ideal of ζ^ω is isomorphic to a chain whose order type is $\Sigma\{\, \zeta^i\omega^* \mid i \in \omega^* \,\}$. These principal ideals are all isomorphic with A_f where f is the transformation of $\omega+1$ where $f(i)=i$ for all $i<\omega$ and $f(\omega)=0$.

REFERENCES

1. A. Clement and F. Pastijn, Inverse semigroups with idempotents dually well-ordered, J. Austral. Math. Soc. (A) 35 (1983), 373-385.
2. J.W. Hogan, Bisimple semigroups with idempotents well-ordered, Semigroup Forum 6 (1973), 298-316.
3. A.C. Morel, A class of relation types isomorphic to the ordinals, Mich. Math. J. 12 (1965), 203-215.
4. W.D. Munn, Fundamental inverse semigroups, Quart. J. Math. Oxford (2) 21 (1970), 157-170.
5. F. Pastijn, Scattered uniform chains, preprint.
6. J.G. Rosenstein, Linear Orderings, Academic Press, New York, 1982.
7. G.L. White, The dual ordinal of a bisimple inverse semigroup, Semigroup Forum 6 (1973), 295-297.

CAYLEY THEOREMS FOR SEMIGROUPS

Mario Petrich
Department of Mathematics
Simon Fraser University
Burnaby, B.C. V5A 1S6
Canada

ABSTRACT. Analogs of the Cayley theorem for groups are considered for various classes of semigroups. First for the class of all semigroups, then for inverse semigroups and for regular semigroups over right normal bands. A problem is proposed on the Cayley type theorems for other classes of semigroups. In particular, a step toward solving this problem for completely regular semigroups is provided by a Cayley theorem for completely simple semigroups.

1. THE RIGHT REGULAR REPRESENTATION

The familiar Cayley theorem for groups is an elementary but far reaching proposition:

Theorem 1. Every group is isomorphic to a permutation group.

This of course is effected by the right regular representation:

$$\rho: s \to \rho_s \qquad (s \in G) \tag{1}$$

where for each $s \in G$,

$$\rho_s: x \to xs \qquad (x \in G). \tag{2}$$

The corresponding statement for semigroups has essentially the same form:

Theorem 2. Every semigroup is isomorphic to a transformation semigroup.

Again the right regular representation is used, this time of S^1, the semigroup S with an identity adjoined if S has none otherwise $S^1 = S$, and then restricted to S. This modification is needed since the right regular representation need not be one-to-one for semigroups.

S. M. Goberstein and P. M. Higgins (eds.), Semigroups and Their Applications, 133–138.

Thus here $\rho: S \to \mathcal{T}(S^1)$, the semigroup of transformations on S^1 (written and composed as right operators). Theorem 2 may be termed "the Cayley theorem for semigroups".

2. REDUCED RIGHT TRANSLATIONS OF WAGNER

An inverse semigroup S has the property that for every $s \in S$, there exists a unique element, denoted by s^{-1}, for which $s = ss^{-1}s$ and $s^{-1} = s^{-1}ss^{-1}$. The mapping ρ_s defined in (2), but now on S, is the (inner) right translation induced by s. This leads us to an important statement:

Theorem 3. (Wagner [2]). Every inverse semigroup is isomorphic to a semigroup of one-to-one partial transformations.

Here the right regular representation is "reduced":

$$\hat{\rho}: s \to \hat{\rho}_s = \rho_s|\{x \in S | x \mathcal{R} xs\} \qquad (s \in S),$$

$\hat{\rho}_s$ is a reduced right translation, giving an isomorphism of S into $\mathcal{I}(S)$, the symmetric inverse semigroup on S. This case is more closely similar to the group case. In an inverse semigroup, we have $Ss^{-1} = \{x \in S | x \mathcal{R} xs\}$, but the above definition may be used in other classes of semigroups.

Indeed, let S be a right normal band, that is a semigroup satisfying the identities $x^2 = x$, $xya = yxa$. A simple argument shows that the mapping $\hat{\rho}$ is an isomorphism of S into the semigroup $\mathcal{F}(S)$ of partial transformations on S. In contradistinction to the three theorems above, this embeds the right normal band S into the semigroup $\mathcal{F}(S)$ which is very far from being a right normal band. This inconvenience may be remedied as follows.

What we call today a right normal band, Wagner called a restrictive semigroup. This is related to the fact that on $\mathcal{F}(X)$ he considered the restricted product defined as follows:

$$\phi \triangleright \psi = \psi|_{\underline{d}\phi}$$

where $\underline{d}\phi$ is the domain of ϕ. Under this product, $\mathcal{F}(X)$ becomes a right normal band and the mapping $\hat{\rho}$ is an isomorphism into it. Fortunately, the restricted product agrees with the usual product (composition) on the image of a right normal band under $\hat{\rho}$. We thus have

Theorem 4. (Wagner [3]). Every right normal band is isomorphic to a semigroup of partial transformations (under the restricted product).

In order to embed a right normal band into a right normal band of partial transformations, we keep the set $\mathcal{F}(X)$ but change its operation. This is a possible alternative to the passage from full to partial transformations where we change the set of transformations but essentially keep their composition.

3. THE MADHAVAN REPRESENTATION

Note that both $\mathcal{T}(X)$ and $\mathcal{I}(X)$ are subsemigroups of $\mathcal{F}(X)$. A further subsemigroup of $\mathcal{F}(X)$ can be constructed as follows. Let θ be an equivalence on a nonempty set X and let

$$\mathcal{F}_\theta(X) = \{\phi \in \mathcal{F}(X) \mid \theta \text{ saturates } \underline{d}\phi,\ \theta|_{\underline{d}\phi} = \underline{e}\phi,\ \theta|_{\underline{r}\phi} = \varepsilon_{\underline{r}\phi}\}$$

where $\underline{d}\phi$ was defined above, $\underline{r}\phi$ is the range of ϕ and $\underline{e}\phi$ is the equivalence on $\underline{d}\phi$ induced by ϕ. Madhavan's system of axioms is different from the above, but they are equivalent. Note that $\mathcal{I}(X) = \mathcal{F}_\varepsilon(X)$ where ε is the equality relation. One shows easily that

$S = \mathcal{F}_\theta(X)$ is closed under composition of partial transformations and

is in fact a <u>regular semigroup over a right normal band</u>. This means that for every $a \in S$, there exists $x \in S$ such that $a = axa$, and its idempotents form a right normal band.

Now let S be a regular semigroup over a right normal band. For each $s \in S$, let

$$V(s) = \{x \in S \mid s = sxs,\ x = xsx\}$$

be the set of all inverses of s. The relation $\mathcal{Y}$ defined by

$$s\ \mathcal{Y}\ t \iff V(x) = V(t) \qquad (s,t \in S)$$

can be proved to be the least inverse semigroup congruence on S.

<u>Theorem</u> 5. (Madhavan [1]). Every regular semigroup over a right normal band is isomorphic to a subsemigroup of $\mathcal{F}_\theta(X)$ for some θ and some X.

In fact, on such a semigroup S we may define

$$\bar{\rho}: s \to \bar{\rho}_s = \rho_{s|(Ss)\mathcal{Y}} \qquad (s \in S)$$

which gives an isomorphism of S into $\mathcal{F}_\mathcal{Y}(S)$. Here

$$(Ss)\mathcal{Y} = \{x \in S \mid x\ \mathcal{Y}\ t \text{ for some } t \in Ss\}$$

is the saturation of Ss by $\mathcal{Y}$.

We can extend the Madhavan representation to include a somewhat larger class of semigroups as follows. Let S be a <u>regular semigroup over a normal band</u> (that is, idempotents satisfy the identity $axya =$

ayxa). Yamada [4] proved that S is a subdirect product of a regular semigroup L over a left normal band and a regular semigroup R over a right normal band. We can represent R as above by means of the Madhavan representation and L by its obvious left-right dual. This gives an isomorphism of S into a direct product of the form $\mathcal{F}'_{\theta'}(X') \times \mathcal{F}_{\theta}(X)$. Note that this product is a regular semigroup over a normal band.

4. ABSTRACTION

From the above examples, we now abstract the following:

Problem 1. Let $\mathcal{C}$ be a class of semigroups. For each nonempty set X construct a family of semigroups of partial transformations $\mathcal{F}_{\mathcal{C}}(X)$ on X, with functions written on the right with a composition not necessarily the usual one. Let $\mathcal{F}'_{\mathcal{C}}(X)$ be the family of semigroups which is obtained by the same construction but writing the functions on the left with the corresponding change in the composition. We further require that every semigroup S in $\mathcal{C}$ admit an isomorphism into a semigroup which is either in $\mathcal{F}_{\mathcal{C}}(X)$ or in $\mathcal{F}'_{\mathcal{C}}(X')$ or is a direct product of a semigroup in $\mathcal{F}_{\mathcal{C}}(X)$ and a semigroup in $\mathcal{F}'_{\mathcal{C}}(X')$ for some sets X and X'.

The other examples exhibit solutions of this problem for groups, semigroups, inverse semigroups, right normal bands and regular semigroups over (right) normal bands. The most intriguing class of semigroups in this context seems to be the following.

A semigroup S is completely regular if it is a union of its (maximal) subgroups. Equivalently, for every $a \in S$ there exists $x \in S$ such that $a = axa$, $ax = xa$. From this we deduce the existence of a^{-1} satisfying $a = aa^{-1}a$, $a^{-1} = a^{-1}aa^{-1}$, $aa^{-1} = a^{-1}a$ (it suffices to take $a^{-1} = xax$). Here a^{-1} is the inverse of a in the maximal subgroup of S containing a.

Much is known about completely regular semigroups. One conspicuous thing missing for them is a (faithful) representation. Also nontrivial examples are lacking as well but this may be a consequence of the lack of a good representation.

Problem 2. Solve Problem 1 for the class of completely regular semigroups.

Toward a solution of this problem, it is reasonable to start with the following special case.

5. A REPRESENTATION OF COMPLETELY SIMPLE SEMIGROUPS

A semigroup S is completely simple if S has no proper ideals and has a primitive idempotent (e is primitive if for any idempotent f, $f = ef = fe$ implies $e = f$).

Let X be a nonempty set and λ be an equivalence relation on X all of whose classes are of the same cardinality and set

$$\mathcal{T}_\lambda(X) = \{\phi \in \mathcal{T}(X) \mid \underline{r}\phi \text{ is a } \lambda\text{-class and for every } \lambda\text{-class } L,\ \phi|_L \text{ is a bijection of } L \text{ onto } \underline{r}\phi\}.$$

One verifies easily that $\mathcal{T}_\lambda(X)$ is a completely simple subsemigroup of $\mathcal{T}(X)$. We can define $\mathcal{T}'_\lambda(X)$ symmetrically obtaining a completely simple subsemigroup of $\mathcal{T}'(X)$.

Theorem 6. For any completely simple semigroup S, the mapping

$$\tau: s \to (\lambda_s, \rho_s) \qquad (s \in S),$$

where $\lambda_s x = sx$ and $x\rho_s = xs$ for all $x \in S$, is a faithful representation

$$S \to \mathcal{T}'_R(S) \times \mathcal{T}_L(S)$$

where the latter is a completely simple semigroup.

This is a combination of the left and the right regular representations. If we take only the right regular representation, we obtain a faithful representaion of a left reductive completely simple semigroup (that is satisfying: $xa = xb$ for all $x \in S$ implies $a = b$). It is instructive to construct a Rees matrix representation of the semigroup $\mathcal{T}_\lambda(X)$.

A completely regular semigroup S always has a congruence η such that S/η is a semilattice (satisfies the identities $x^2 = x$, $xy = yx$) and all η-classes are completely simple semigroups. This in fact characterizes completely regular semigroups and there are several constructions of them based on this fact. It should by now be clear that Problem 2 is very far from being solved.

6. Besides completely regular semigroups, there are many other interesting classes of semigroups for which a solution of Problem 1 may have important consequences. Among these, one may take a number of varieties of completely regular semigroups which then would provide natural examples of semigroups belonging to these varieties.

REFERENCES

[1] S. Madhavan, 'On right normal right inverse semigroups', Semigroup Forum 12 (1976), 333-339.

[2] V.V. Wagner, 'Generalized groups', Doklady Akad. Nauk SSSR 84 (1952), 1119-1122 (Russian).

[3] V.V. Wagner, 'Restrictive semigroups', Izv. vysš. učebn. zav. Matem. No. 6 (31) (1962), 19-27 (Russian).

[4] M. Yamada, 'Regular semigroups whose idempotents satisfy permutation identities', Pacific J. Math. 21 (1967), 371-392.

POWER SEMIGROUPS AND RELATED VARIETIES OF FINITE SEMIGROUPS

J.E. PIN
C.N.R.S, L.I.T.P. Université Paris VI
Tour 55-65
4 place Jussieu
75252 Paris Cedex 05

As the title suggests, all semigroups considered in this paper will be finite. Let S be a semigroup. The power semigroup (or "global") of S, $\mathcal{P}(S)$, is the set of all subsets of S with multiplication defined, for all $X, Y \in S$ by

$$XY = \{xy \mid x \in X, y \in Y\}$$

A number of papers has been devoted to the study of power semigroups, especially in connection with the following problem of Schein : if $\mathcal{P}(S)$ is isomorphic to $\mathcal{P}(T)$, is S isomorphic to T ? We shall not discuss this challenging problem in this survey, although some of the results mentioned below are related to this question, but we shall focus our interest on "structural" properties of power semigroups (idempotents, groups, Green relations, etc.) and on varieties generated by power semigroups. Only a very few (elementary) proofs will be given.

1. EXAMPLES AND ELEMENTARY PROPERTIES

We first observe that the empty set is always a zero of $\mathcal{P}(S)$. Therefore it is natural to consider also the semigroup $\mathcal{P}'(S)$ of all non-empty subsets of S. Then S is a subsemigroup of $\mathcal{P}'(S)$ and $\mathcal{P}'(S)$ is a subsemigroup of $\mathcal{P}(S)$. Furthermore $\mathcal{P}(S)$ is a quotient of $\mathcal{P}'(S) \times U_1$ where U_1 denotes the monoid consisting of an identity and of a zero.

Let us first give two examples of power semigroups

Example 1

Let $S = \{a,b,0\}$ be the semigroup presented on $\{a,b\}$ by the relations $a^2 = a$, $ab = b$ and $ba = b^2 = 0$. Then the structure of S and $\mathcal{P}(S)$ can be represented as follows (as usual, a star indicates an idempotent in the corresponding $\mathcal{H}$-class, an "egg-box" represents a $\mathcal{D}$-class (= $\mathcal{J}$-class), which might be divided into $\mathcal{R}$-classes and $\mathcal{L}$-classes, and the order is the $\mathcal{J}$-ordering: $a \leq_{\mathcal{J}} b$ iff the ideal generated by a

S. M. Goberstein and P. M. Higgins (eds.), Semigroups and Their Applications, 139–152.

is contained in the ideal generated by b.)

S :

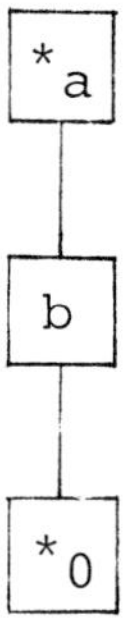

$\mathcal{P}(S)$

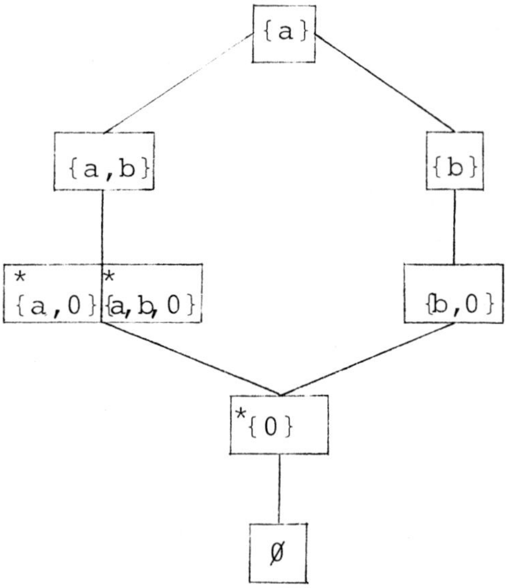

Example 2

Let $S = BA_2$ be the 5-element Brandt aperiodic semigroup ("The universal counter-example"). Then S has the following structure

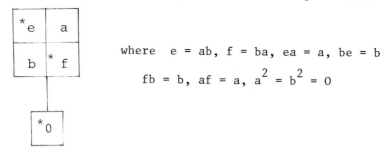

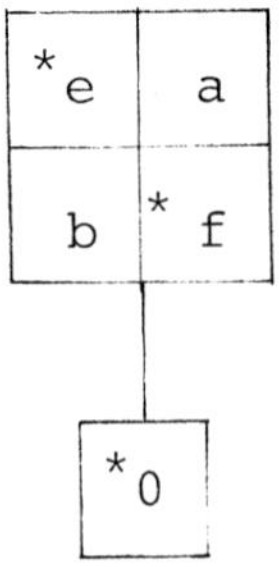

where $e = ab$, $f = ba$, $ea = a$, $be = b$

$fb = b$, $af = a$, $a^2 = b^2 = 0$

One can verify that, in $\mathcal{P}(S)$, $\{e,b\}\ \mathcal{R}\ \{a,f\}$, $\{a,e\}\ \mathcal{L}\ \{b,f\}$, and $\{\{a,b,o\}, \{e,f,o\}\}$ forms a group isomorphic to $\mathbb{Z}/2\mathbb{Z}$.

More generally, we denote by BA_n the semigroup of boolean n x n matrices having at most one non zero entry and by M_n the semigroup of all boolean n x n matrices. Thus, $BA_2 = \{a,b,e,f,o\}$ where

$a = \begin{pmatrix} 0 & 1 \\ 0 & 0 \end{pmatrix}$ $b = \begin{pmatrix} 0 & 0 \\ 1 & 0 \end{pmatrix}$ $e = \begin{pmatrix} 1 & 0 \\ 0 & 0 \end{pmatrix}$ $f = \begin{pmatrix} 0 & 0 \\ 0 & 1 \end{pmatrix}$ and $0 = \begin{pmatrix} 0 & 0 \\ 0 & 0 \end{pmatrix}$

Then we have

<u>Proposition 1.1</u> M_n is a quotient of $\mathcal{P}(BA_n)$.

<u>Proof</u> Associate to any subset X of BA_n the matrix $\sum_{x \in X} x$. This defines a surjective morphism from $\mathcal{P}(BA_n)$ onto M_n.

<u>Corollary 1.2</u> Every semigroup divides $\mathcal{P}(BA_n)$ for some $n > 0$.

Properties of semigroups are not usually inherited by power semigroups. However this is the case for some special properties. Recall that a semigroup S is left (right) trivial if, for every idempotent $e \in S$, $eS = e$ (resp. $Se = e$). Note that a semigroup that is both left and right trivial is nilpotent. S is locally trivial if, for every idempotent $e \in S$, $eSe = e$.

<u>Proposition 1.3</u>

(a) If S is a commutative semigroup, so are $\mathcal{P}'(S)$ and $\mathcal{P}(S)$.

(b) More generally, if S satisfies a permutative identity $x_1 \ldots x_n = x_{1\sigma} \ldots x_{n\sigma}$, so do $\mathcal{P}'(S)$ and $\mathcal{P}(S)$.

<u>Proposition 1.4</u>

(a) If S is a left (right) zero semigroup, so is $\mathcal{P}'(S)$.

(b) If S is a left trivial (right trivial, nilpotent) semigroup, so is $\mathcal{P}'(S)$.

(c) If S is a locally trivial semigroup, so is $\mathcal{P}'(S)$.

Note that proposition 1.4 is false for $\mathcal{P}(S)$: consider $\mathcal{P}(1) = U_1 \ldots$

<u>Corollary 1.5</u> Let S be a rectangular band. Then $\mathcal{P}'(S)$ is a band if and only if S is a left or right zero semigroup.

One of the more interesting examples of power semigroups is the power semigroup of a group. This will be studied in the next section.

2. POWER SEMIGROUP OF A GROUP

We first describe the idempotents.

Proposition 2.1 Let G be a group. Then the idempotents of $\mathcal{P}'(G)$ are the subgroups of G.

Green relations also admit a simple description.

Proposition 2.2 Let G be a group and let $X, Y \in \mathcal{P}'(G)$.

(a) $X \mathcal{R} Y$ (resp. $X \mathcal{L} Y$) if and only if there exists $g \in G$ such that $Xg = Y$ (resp. $gX = Y$).

(b) $X \mathcal{D} Y$ if and only if there exist $g_1, g_2 \in G$ such that $g_1Xg_2 = Y$. In particular if $X \mathcal{D} Y$ then $|X| = |Y|$.

Corollary 2.3 Let H and K be two subgroups of a group G. Then H and K are $\mathcal{D}$- equivalent in $\mathcal{P}'(G)$ if and only if they are conjugate.

We can now completely describe the regular $\mathcal{D}$-classes of $\mathcal{P}'G)$. Recall that the normalizer of a subgroup H of G is the group $N(H) = \{g \in G \mid gH = Hg\}$.

Theorem 2.4 Let H be a subgroup of a group G and let D be the $\mathcal{D}$-class of $\mathcal{P}'(G)$ containing H. Then D is a Brandt $\mathcal{D}$-class of size $|G : N(H)|$ with structure group $N(H)/H$.

* H	Hg_1	Hg_2	Hg_3	Hg_4
h_1H	*			
h_2H		*		
h_3H			*	
h_4H				*

3. STRUCTURE PROPERTIES OF POWER SEMIGROUPS

In this section, we come back to the general case, and try to relate properties of S and of $\mathcal{P}(S)$. We first describe the idempotents of $\mathcal{P}(S)$.

Proposition 3.1 Let S be a semigroup and let T be an element of $\mathcal{P}(S)$. Then T is idempotent if and only if T is a subsemigroup of S whose maximal $\mathcal{D}$-classes are regular.

As shown in example 2, the power semigroup of an aperiodic semigroup needs not be aperiodic. However, we have the following nice result, due to Putcha.

Theorem 3.2 [30, 26] Let S be an aperiodic semigroup. The following conditions are equivalent.

(1) $\mathcal{P}(S)$ is aperiodic.

(2) BA_2 does not divide S.

(3) For all idempotents e,f in S, $e \mathcal{D} f$ implies $e \mathcal{D} ef$ or $e \mathcal{D} fe$.

Condition (3) expresses the fact that, in a regular $\mathcal{D}$-class of S, only the patterns 2,3 or 4 are allowed

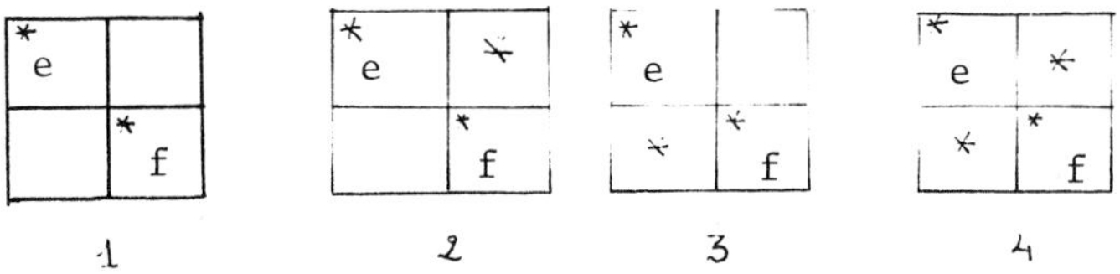

Theorem 3.2 characterizes the semigroups S such that $\mathcal{P}(S)$ is $\mathcal{H}$-trivial. Similar results hold for the other Green relations. We denote by $E(S)$ the set of idempotents of S.

Theorem 3.3 [26] Let S be an aperiodic semigroup. Then

(a) $\mathcal{P}(S)$ is $\mathcal{R}$-trivial if and only if, for every $s \in S$ and $e \in E(S)$, $es = ese$.

(b) $\mathcal{P}(S)$ is $\mathcal{L}$-trivial if and only if, for every $s \in S$ and $e \in E(S)$, $se = ese$.

(c) $\mathcal{P}(S)$ is $\mathcal{J}$-trivial if and only if, for every $s \in S$ and $e \in E(S)$, $es = se$.

Theorem 3.2 can be extended in another direction.

Theorem 3.4 [15] Let S be a semigroup such that BA_2 does not divide S. Then every group in $\mathcal{P}(S)$ divides a direct product of groups in S.

Corollary 3.5 Let S be a semigroup all of whose regular $\mathcal{D}$-classes are simple semigroups. Then every group in $\mathcal{P}(S)$ divides a direct product of groups of S.

Note that the study of subgroups of a power semigroup is not yet achieved. The first step would be a complete description of the groups in $\mathcal{P}(S)$ when S is a 0-simple semigroup.

4. VARIETIES GENERATED BY POWER MONOIDS

Recall that a variety of finite monoids (or "pseudovariety") is a class of finite monoids closed under taking submonoids, quotients and finite direct products. Given a variety $\underline{V}$ of (finite) monoids, we denote by $\underline{PV}$ the smallest variety containing all the monoids of the form $\mathcal{P}(M)$ where $M \in \underline{V}$. This defines an operator $\underline{P}$ on varieties that can be iterated by setting, for every variety $\underline{V}$,

$$\underline{P}^{\circ}\underline{V} = \underline{V} \qquad \underline{P}^{n+1}\underline{V} = \underline{P}(\underline{P}^{n}\underline{V})$$

We first give a classification of varieties of the form $\underline{PV}$ when $\underline{V}$ is a commutative variety, that is, a variety that only contains commutative monoids. Let us start with the trivial variety, containing the monoid 1 only.

Proposition 4.1 Let $\underline{V}$ be the trivial variety. Then

(a) $\underline{PV}$ is the variety of the idempotent and commutative monoids (semilattices).

(b) $\underline{P}^2\underline{V}$ is the variety of the aperiodic and commutative monoids.

(c) $\underline{P}^3\underline{V} = \underline{P}^2\underline{V}$.

If $\underline{V}$ is non trivial, we have the following result

Proposition 4.2 [24,32,25] Let $\underline{V}$ be a non trivial commutative variety, and let $\underline{H}$ be the class of all the groups that are subgroups of some $M \in \underline{V}$. Then

(a) $\underline{PV}$ consists of the commutative monoids whose subgroups are elements of $\underline{H}$.

(b) $\underline{P}^2\underline{V} = \underline{PV}$

Note that $\underline{H}$ is in fact a variety of monoids, consisting of commutative groups. For instance, let $\underline{V}$ be the variety of idempotent and commutative monoids. Then $\underline{H}$ is the trivial variety (since every subgroup of a semilattice is trivial) and therefore $\underline{PV}$ is the variety of the commutative monoids whose subgroups are trivial : this is just proposition 4.1.b).

The non-commutative case is much more involved. However, as in the commutative case, the sequence $\underline{P}^n\underline{V}$ is finite.

Theorem 4.3 [16] Let $\underline{V}$ be a variety containing at least one non commutative monoid. Then $\underline{P}^3\underline{V}$ is the variety $\underline{M}$ of all the monoids and therefore $\underline{P}^3\underline{V} = \underline{P}^4\underline{V}$.

The next result shows that the bound 3 is the best possible.

Theorem 4.4 [25] Let $\underline{V}$ be the variety of the $\mathcal{R}$-trivial and idempotent monoids, defined by the equations $x = x^2$ and $xy = xyx$. Then

$$\underline{V} \subsetneq \underline{PV} \subsetneq \underline{P}^2\underline{V} \subsetneq \underline{P}^3\underline{V} = \underline{M}.$$

These results suggest a first classification of non commutative varieties. Define the exponent of a non commutative variety as the smallest integer n such that $\underline{P}^n\underline{V} = \underline{M}$. By theorem 4.3, the exponent is always 0,1, 2 or 3. Clearly, $\underline{M}$ is the only variety of exponent 0, and the next theorem gives a characterization of the varieties of exponent 1. In this statement, $\underline{DS}$ denotes the variety of the monoids whose regular $\underline{D}$-classes are $\underline{S}$emigroups.

Theorem 4.5 [14] Let $\underline{V}$ be a non commutative variety. The following conditions are equivalent.

(i) $\underline{PV} = \underline{M}$ (the exponent of $\underline{V}$ is 0 or 1).

(ii) $BA_2 \in \underline{V}$.

(iii) $\underline{V}$ is not contained in $\underline{DS}$.

Maybe it is more convenient to restate this result as follows.

Corollary 4.6 Every variety of exponent 2 or 3 is contained in $\underline{DS}$.

At this point it remains to separate the varieties of exponent 2 from those of exponent 3. Here is a first step in this direction.

Theorem 4.7 [25] If $\underline{V}$ contains a non commutative group, then $\underline{P}^2\underline{V} = \underline{M}$.

This means that every group contained in a variety of exponent 3 is commutative. Corollary 3.6 also suggests the existence of a variety containing all the varieties of exponent 3. Unfortunately, the situation is more involved, and not yet completely solved. However, one can characterize the aperiodic varieties of exponent 3.

Let $\mathbf{V}_1$ (resp. $\mathbf{V}_2$) be the variety of aperiodic monoids M such that, for every $s \in M$ and $e \in E(M)$, $es = ese$ (resp. $se = ese$). One can show that $\mathbf{V}_1$ (resp. $\mathbf{V}_2$) is the variety generated by all the monoids of the form S^1 where S is a left trivial (right trivial) semigroup.

Let $\mathbf{V}_3$ be the variety defined by the equations $x = x^2$ and $xyxzx = xyzx$. $\mathbf{V}_3$ is the variety generated by the monoids U_2 and U_2^r

U_2 : 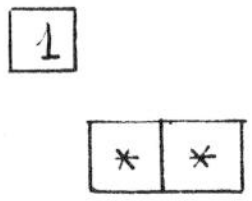

U_2^r : 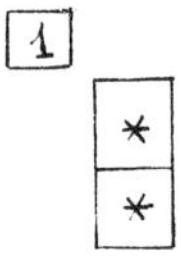

or by the monoid U

U : 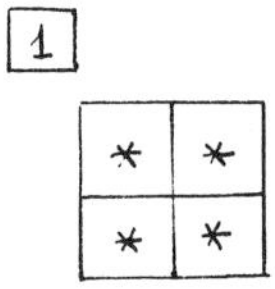

We can now state

Theorem 4.8 Let $\mathbf{V}$ be a non commutative aperiodic variety. Then $\mathbf{V}$ has exponent 3 (i.e. $\mathbf{V} \subsetneq \mathbf{PV} \subsetneq \mathbf{P}^2\mathbf{V} \subsetneq \mathbf{P}^3\mathbf{V} = \mathbf{M}$) if and only if $\mathbf{V}$ is contained in one of the three varieties $\mathbf{V}_1$, $\mathbf{V}_2$ or $\mathbf{V}_3$.

The varieties $\mathbf{V}_1$ and $\mathbf{V}_2$ also play a role in a different context. Set $\mathbf{V}_0 = \mathbf{V}_1 \cap \mathbf{V}_2$. Thus $\mathbf{V}_0$ is the variety of the aperiodic monoids such that, for every $s \in M$ and $e \in E(M)$, $es = se$.

Theorem 4.9 [26] Let $\mathbf{V}$ be a variety of monoids. Then

(a) $\mathbf{PV} = \mathbf{J}$ if and only if $\mathbf{V}$ is a non commutative variety contained in $\mathbf{V}_0$.

(b) $\mathbf{PV} \subset \mathbf{R}$ if and only if $\mathbf{V} \subset \mathbf{V}_1$.

(c) $\mathbf{PV} \subset \mathbf{R}^r$ if and only if $\mathbf{V} \subset \mathbf{V}_2$.

It is an open problem to know whether $\mathbf{PV}_1 = \mathbf{R}$. The next iteration of the operator $\mathbf{P}$ produces the important variety $\mathbf{PJ}$. Let $\mathbf{DA}$ be the variety of monoids whose regular D-classes are Aperiodic semigroups (in fact, restangular bands!) and let V_4 be the variety of $\mathcal{J}$-trivial monoid with commuting idempotents.

Theorem 4.10 [28] For every variety V such that $V_4 \subset V \subset DA$, PV = PJ.

In particular $PV_4 = PJ = PR = PR^r = PDA$.

It is not yet known whether the variety PJ is decidable (A variety V is decidable if there exists an algorithm to decide whether a given monoid belongs to V or not). In fact this problem is equivalent to a famous problem of language theory, the "dot-depth 2" problem. (See [28] and the recent work of Straubing for a discussion.)

The results of section 2 can also be interpreted in terms of varieties. Denote by G the variety of groups, and by BG the variety of all the monoids whose regular $\mathcal{D}$-classes are Brandt (BG stands for "block groups"). Then theorem 2.4. can be formulated as follows.

Proposition 4.11 [18] PG is contained in BG.

A number of descriptions of PG and BG are known (see [18] for more details) but the main question remains open : is PG equal to BG? If the answer is "yes", this would prove that PG is decidable, but there are some reasons to think that the inclusion might be strict. On the other hand, if PG is strictly contained in BG, PG should ultimately satisfy some equations that are not satisfied by BG. It would be very interesting to find out these mysterious equations !

These problems show that the classification of the varieties of the form PV is far to be achieved. We even don't know if there are infinitely many aperiodic varieties of the form PV...

Of course, the same problems arise for varieties of semigroups and the situation is even more complicated ! However, Almeida [1,2, and forthcoming papers] has characterized the varieties of semigroups V such that PV = V and has completely described the restriction of the operator P' (associated to $\mathcal{P}'(S)$) to the varieties of semigroups contained in LI, the variety of locally trivial semigroups : for these varieties, the equality $P'V = P'^2V$ always hold. He has also given a description of P^2V when V is permutative and has shown that, if V is non permutative, then $P^3V = S$, the variety of all the semigroups. This extends theorem 4.3 to the semigroup case.

5. CONNECTIONS WITH LANGUAGE THEORY

Language theory has not only motivated the study of power semigroups, but has also permitted to prove some of the results mentioned before, for instance theorems 4.3, 4.4, 4.9, 4.10. We refer the reader to [27, E, L] for undefined notations of this section.

Let A be a finite alphabet. We denote by A^+, the free semigroup on A. Recall that a semigroup S recognizes a language L of A^+ if there exists a semigroup morphism $\eta : A^+ \to S$ and a subset P of S such that $L = P\eta^{-1}$. The key result is the following

<u>Proposition 5.1</u> Let $L \subset A^+$ be a language and let $\varphi : A^+ \to B^+$ be a length-preserving morphism. If L is recognized by S, then $L\varphi$ is recognized by $\mathcal{P}(S)$.

<u>Proof</u> By assumption, there exists a semigroup morphism $\eta : A^+ \to S$ and a subset P of S such that $L = P\eta^{-1}$. Set $R = \{ Q \in \mathcal{P}(S) \mid Q \cap P \neq \emptyset \}$ and define a morphism $\pi : B^+ \to \mathcal{P}(S)$ by setting $b\pi = (b\varphi^{-1})\eta$. Then one can verify that $R\pi^{-1} = L\varphi$ and thus $\mathcal{P}(S)$ recognizes $L\varphi$.

<u>Corollary 5.2</u> Let S be the syntactic semigroup of L. Then the syntactic semigroup of $L\varphi$ divides $\mathcal{P}(S)$.

Although this corollary is very elementary, it can be used to avoid some tedious computations. The basic idea is that it is usually much more easy to compute a language of the form $L\varphi$ than the power semigroup of a semigroup S. As an illustration, we shall prove again that $\mathbb{Z}/2\mathbb{Z}$ divides $\mathcal{P}(BA_2)$ (cf. example 2).

Let $A = \{a,b\}$. As well known, BA_2 is the syntactic semigroup of the language $(ab)^+$. We define $\varphi : \{a,b\}^+ \to a^+$ by $a\varphi = b\varphi = a$. Obviously $(ab)^+\varphi = (a^2)^+$ and by corollary 5.2, the syntactic semigroup of $(a^2)^+$ - which is precisely $\mathbb{Z}/2\mathbb{Z}$ - divides $\mathcal{P}(BA_2)$.

However the more important result is a form of converse to proposition 5.1. Recall that a variety of languages $\mathcal{V}$ associates to every alphabet A, a class $A^+\mathcal{V}$ of recognizable languages of A^+ such that:

(a) For every alphabet A, $A^+\mathcal{V}$ is a boolean algebra.

(b) For every semigroup morphism $\varphi : A^+ \to B^+$, $L \in B^+\mathcal{V}$ implies $L\varphi^{-1} \in A^+\mathcal{V}$.

(c) For every $a \in A$ and for every $L \in A^+\mathcal{V}$, $a^{-1}L$, $La^{-1} \in A^+\mathcal{V}$.

Let $\underline{V}$ be a variety of semigroups. For each alphabet A, we denote by $A^+\mathcal{V}$ the set of all the languages recognized by a semigroup of $\underline{V}$; $\mathcal{V}$ is the variety of languages corresponding to $\underline{V}$. Then we have

<u>Theorem 5.2</u> [31, 32]. Let $\underline{V}$ be a variety of semigroups and let $\mathcal{V}$ (resp. $\mathcal{W}$) be the variety of languages corresponsing to $\underline{V}$ (resp. $\underline{PV}$). Then, for every alphabet A, $A^+\mathcal{W}$ is the boolean algebra generated by

all the languages of the form $L\varphi$, where $L \in B^+\mathcal{V}$ and $\varphi : B^+ \to A^+$ is a length-preserving morphism.

Now, by Eilenberg's theorem, the correspondence $\underline{V} \longrightarrow \mathcal{V}$ is one-to-one, and language theoretic methods can be used to study $\underline{PV}$... See [16, 17, 18, 25, 28].

Power semigroups are also very important in the study of relations from A^* into a monoid M, but this subject is so wide that another article would not suffice...

Acknowledgements. I would like to thank J. Almeida and S.W. Margolis for many helpful contributions in the preparation of this survey.

REFERENCES

N.B. These references are yet far to be complete, especially for the earlier references, and the author would appreciate any new reference to add to this list.

[1] J. Almeida, 'On power varieties of semigroups', to appear.

[2] J. Almeida, 'Power pseudovarieties of semigroups. I, II', to appear.

[3] S.G. Beršadskiĭ, 'Embeddability of semigroups in a global super-semigroup of a group', Semigroup Varieties and Semigroups of Endomorphisms, Leningrad. Gos. Ped. Inst. Leningrad (1979) 47-49 (Russian).

[4] C. Fox and J. Rhodes, 'The complexity of the power set of a semigroup'. Preprint, University of California, Berkeley (1984).

[5] M. Gould and J.A. Iskra, 'Globally determined classes of semigroups', Semigroup Forum.

[6] M. Gould and J.A. Iskra, 'Embedding in globals of groups and semilattices', to appear.

[7] M. Gould, J.A. Iskra and P.P. Pálfy, 'Embedding in global of finite semilattices', preprint.

[8] M. Gould, J.A. Iskra and C. Tsinakis, 'Globally determined lattices and semilattices', Algebra Universalis.

[9] M. Gould, J.A. Iskra and C. Tsinakis, 'Globals of completely regular periodic semigroups', to appear.

[10] Y. Kobayashi, 'Semilattices are globally determined', Semigroup Forum 29 (1984) 217-222.

[11] A. Lau, 'Finite abelian semigroups represented into the power set of finite groups', Czech. Math. J., 29 (1979), 159-162.

[12] E.S. Ljapin, 'Identities valid globally in semigroups', Semigroup Forum 24 (1982) 263-269.

[13] D.J. McCarthy and D.L. Hayes, 'Subgroups of the power semigroup of a group', Journal of Combinatorial Theory (A) 14 (1973) 173-186.

[14] S.W. Margolis, 'On M-varieties generated by power monoids', Semigroup Forum 22 (1981) 339-353.

[15] S.W. Margolis, unpublished.

[16] S.W. Margolis and J.E. Pin, 'Minimal non commutative varieties and power varieties', Pacific Journal of Mathematics 111 (1984) 125-135.

[17] S.W. Margolis and J.E. Pin, 'Power monoids and finite J-trivial monoids', Semigroup Forum 29 (1984) 99-108.

[18] S.W. Margolis and J.E. Pin, 'Varieties of finite monoids and topology for the free monoid', Proceedings of the Marquette Conference on Semigroups, Milwaukee (1984).

[19] E.M. Mogiljanskaja, 'Global definability of certain semigroups', Uč. Zap. Leningrad. Gos. Ped. Inst. 404 (1971) 146-149 (Russian).

[20] E.M. Mogiljanskaja, 'On the definability of certain idempotent semigroups by the semigroup of their subsemigroups', Uč. Zap. Leningrad. Gos. Ped. Inst. 496 (1972), 37-48 (Russian).

[21] E.M. Mogiljanskaja, 'Definability of certain holoid semigroups by means of the semigroups of all their subsets and subsemigroups', Uč. Zap. Leningrad. Gos. Ped. Inst. 496 (1972), 49-60 (Russian).

[22] E.M. Mogiljanskaja, 'The solution of a problem of Tamura', Sbornik Naučnyh Trudov Leningrad. Gos. Ped. Inst. "Modern Analysis and Geometry" (1972), 148-151. (Russian).

[23] E.M. Mogiljanskaja, 'Non isomorphic semigroups with isomorphic semigroups of subsets', Semigroup Forum 6 (1973), 330-333.

[24] J.F. Perrot, 'Varietés de langages et opérations', Theoretical Computer Science 7(1978) 197-210.

[25] J.E. Pin, 'Varietés de langages et monoïde des parties', Semigroup Forum 20 (1980) 11-47.

[26] J.E. Pin, 'Semigroupe des parties et relations de Green', Can. J. Math. 36 (1984) 327-343.

[27] J.E. Pin, Varieties of formal languages. Masson, Paris (1984), English translation (1986), North Oxford Academy.

[28] J. E. Pin and H. Straubing, 'Monoids of upper triangular matrices', Colloquia Mathematica Societatis Janos Bolyai (1981), 259-272.

[29] M.S. Putcha, 'On the maximal semilattice decomposition of the power semigroup of a semigroup', Semigroup Forum 15 (1978) 263-267.

[30] M.S. Putcha, 'Subgroups of the power semigroup of a finite semigroup', Can. J. Math. 31 (1979) 1077-1083.

[31] Ch. Reutenauer, 'Sur les varietés de langages et de monoïdes', 4th GI Conference Lect. Notes in Comp. Sc. 67, Springer, (1979), 260-265.

[32] H. Straubing, 'Recognizable sets and power sets of finite semigroups', Semigroup Forum 18 (1979) 331-340.

[33] T. Tamura, 'The power semigroups of rectangular groups and chains', preprint.

[34] T. Tamura, 'Power semigroups of completely simple semigroups', preprint

[35] T. Tamura and J. Shafer, 'Power semigroups', Math. Japon, 12 (1967), 25-32.

[36] T. Tamura and J. Shafer, 'Power semigroups', Notices AMS 14 (1967) 688.

[37] T. Tamura and J. Shafer, 'Power semigroups II', Notices AMS 15 (1968), 395.

References on languages

[E] S. Eilenberg, Automata, Languages and Machines, Academic Press, Vol. A (1974), Vol. B (1976)

[L] G. Lallement, Semigroups and Combinatorial Applications, Wiley (1979)

ON THE LATTICE OF VARIETIES OF COMPLETELY REGULAR SEMIGROUPS

Norman R. Reilly
Department of Mathematics
Simon Fraser University
British Columbia, Canada, V5A 1S6

ABSTRACT. Completely regular semigroups are semigroups which are (disjoint) unions of groups. In this article we review the progress made in the study of the lattice $L(CR)$ of varieties of completely regular semigroups. Various results describing the lattice of subvarieties $L(U)$ of some proper subvariety U of CR are given; for example, with U equal to the variety of bands, central completely simple semigroups and several other interesting varieties. In the final section, certain global decompositions of $L(CR)$ using the concepts of kernel, left trace and right trace are discussed.

1. INTRODUCTION

The class of completely regular semigroups (semigroups that are unions of groups) has attracted the attention of mathematicians ever since the pioneer work of Rees [45] and Clifford [2] in the early 1940's. As unions of such familiar objects as groups they are natural objects to investigate. The simple fact of being a union of groups would suggest that such objects should be simple to investigate (modulo group theory). This impression was perhaps reinforced by the first elegant and simple structure theorems of Rees and Clifford. However, this is far from the case and their structure still remains rather mysterious in many respects. Numerous authors have attempted to illustrate the structure of arbitrary completely regular semigroups (Clifford [2], Clifford and Petrich [6], Petrich [30], [34], Warne [49] and Yamada [51]) and for specific classes (Clifford [3], [4], [5] Clifford and Petrich [6], Fantham [7], Gerhard and Petrich [16], Green and Rees [17], Jones [19], Kadourek and Polák [21], Leech [23], Petrich [30], Rees [45], Trotter [47], [48], Warne [49] and Yamada [51]).

Commencing with the work of Birjukov [1], Fennemore [9] and Gerhard [10] on varieties of bands there has been a growing interest in the study of varieties of completely regular semigroups.

The basic examples of completely regular semigroups are (i) groups, (ii) bands and (iii) completely simple semigroups.

S. M. Goberstein and P. M. Higgins (eds.), Semigroups and Their Applications, 153–167.

A semigroup S is said to be a <u>semilattice of semigroups</u> if
(i) there exists a semilattice Y and
(ii) for each $\alpha \in Y$ there is a semigroup S_α such that
(iii) $S = \bigcup_{\alpha \in Y} S_\alpha$ (the disjoint union)
(iv) for all $\alpha, \beta \in Y$, $S_\alpha S_\beta \subseteq S_{\alpha\beta}$.

<u>Theorem</u> 1.1 (Clifford [2]). A semigroup S is completely regular if and only if S is a semilattice of completely simple semigroups.

If we consider completely regular semigroups as algebras with a binary operation (multiplication) and a unary operation $x \to x^{-1}$ (the inverse of x in the group containing it), then the class $\mathcal{CR}$ of all completely regular semigroups becomes a variety:

$$\mathcal{CR} = [x = xx^{-1}x,\ xx^{-1} = x^{-1}x,\ (x^{-1})^{-1} = x] .$$

We shall write $x^0 = xx^{-1} = x^{-1}x$. These defining identities for $\mathcal{CR}$ are assumed to be present throughout the discussions below. This view of the class of completely regular semigroups as a variety was introduced by Petrich [31], [32]. A discussion of many important subvarieties of $\mathcal{CR}$ and their defining identities can be found in [33].

The set $L(\mathcal{CR})$ of all varieties of completely regular semigroups forms a lattice with respect to the operations $\vee, \wedge$:

$$\mathcal{U} \wedge \mathcal{V} = \mathcal{U} \cap \mathcal{V},\quad \mathcal{U} \vee \mathcal{V} = \cap\{\mathcal{W} : \mathcal{W} \in L(\mathcal{CR}),\ \mathcal{U},\mathcal{V} \subseteq \mathcal{W}\}.$$

For any $\mathcal{U} \in L(\mathcal{CR})$, $L(\mathcal{U})$ will denote the lattice of all subvarieties of $\mathcal{U}$.

2. BANDS

Certain important sublattices of $L(\mathcal{CR})$ have been studied in great detail. The best understood of these is the lattice of varieties of bands. The variety $\mathcal{B}$ of bands is defined by the simple identity $x^2 = x$. The lattice $L(\mathcal{B})$ was determined independently by Birjukov [1], Fennemore [9] and Gerhard [10].

A great deal is known about $\mathcal{B}$. For instance, descriptions of the free objects in the various subvarieties can be obtained from descriptions of the free band (Green and Rees [17], Gerhard and Petrich [16]). The subdirectly irreducible bands have also been characterized (Gerhard [11]).

A related concept is that of band monoids where, in addition to the usual multiplication, there is a nullary operation to yield the identity. Wismath [50] has shown that the mapping

$$\text{Mon} : L(\mathcal{B}) \to L(\mathcal{BM})$$

where $\mathcal{BM}$ is the variety of band monoids and Mon $(\mathcal{U}) = \{S : S \in \mathcal{U}$ and $S = S^1\}$ is an epimorphism of $L(\mathcal{B})$ onto $L(\mathcal{BM})$. The congruence induced on $L(\mathcal{B})$ by Mon is illustrated in Figure 1a, where Mon-equivalent elements are connected by solid lines, and the image is given in Figure 1b.

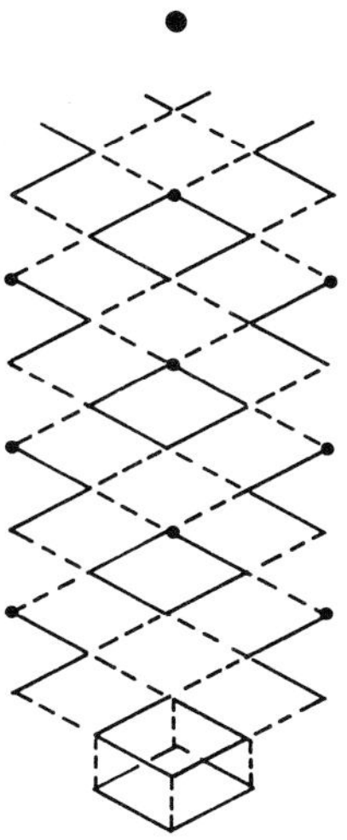

Figure 1a

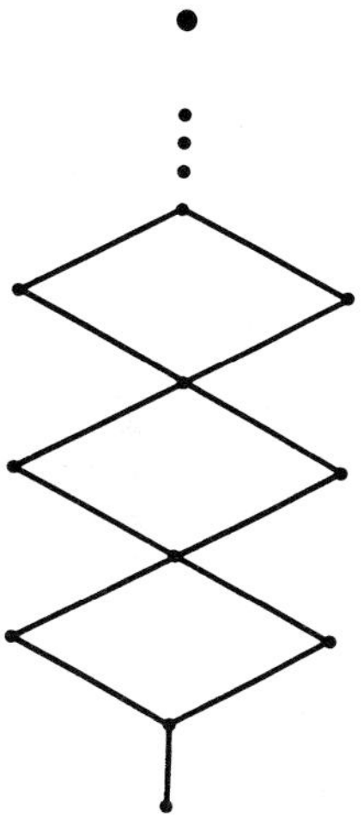

$L(\mathcal{BM})$

Figure 1b

3. ORTHODOX BANDS OF GROUPS

As a counterpoint to the variety $\mathcal{B}$ of bands we have the variety $\mathcal{G}$ of groups. Clearly $\mathcal{G}$ and $\mathcal{B}$ have only the trivial variety $\mathcal{T}$ in common: $\mathcal{G} \cap \mathcal{B} = \mathcal{T}$. The lattice of subvarieties of $\mathcal{G}$ is extremely complicated and has been extensively studied. A great deal of information about $L(\mathcal{G})$ can be found in Neumann [25] and, updated, in Kovács and Newman [22].

A natural step beyond $L(\mathcal{G})$ and $L(\mathcal{B})$ was the investigation of $L(\mathcal{G} \vee \mathcal{B})$. This was successfully accomplished by M. Petrich [31]. It is clear that for any $S \in \mathcal{G} \cup \mathcal{B}$, the set of idempotents $E(S)$ of S is a subsemigroup of S, that is, S is orthodox. This is characterized by the identity $x^0y^0 = (x^0y^0)^0$, and we denote by $\mathcal{O}$ the variety of all orthodox completely regular semigroups: $\mathcal{O} = [x^0y^0 = (x^0y^0)^0]$. Any $S \in \mathcal{G} \cup \mathcal{B}$ also has the property that Green's relation $\mathcal{H}$ is a congruence: such a semigroup is called a band of groups. This is characterized by the identity $(x^0y^0)^0 = (xy)^0$. We denote by $\mathcal{BG}$ the variety of all bands of groups: $\mathcal{BG} = [(x^0y^0)^0 = (xy)^0]$. Thus we have $\mathcal{G} \vee \mathcal{B} \subseteq \mathcal{O} \cap \mathcal{BG}$. In fact (Petrich [31]), we have equality: $\mathcal{G} \vee \mathcal{B} = \mathcal{O} \cap \mathcal{BG}$ and this variety is denoted by $\mathcal{OBG}$ - the variety of orthodox bands of groups.

The lattice $L(\mathcal{OBG})$ has been described in terms of $L(\mathcal{B})$ and $L(\mathcal{G})$ as follows:

<u>Theorem</u> 3.1 (Petrich [31]). The mapping

$$\chi : \mathcal{W} \to (\mathcal{W} \cap \mathcal{B}\ ,\ \mathcal{W} \cap \mathcal{G})$$

is an isomorphism of $L(\mathcal{OBG})$ onto $L(\mathcal{B}) \times L(\mathcal{G})$. If $\mathcal{W}\chi = (\mathcal{U},\mathcal{V})$, then

$$\mathcal{W} = \mathcal{U} \vee \mathcal{V} \qquad (1)$$

$$= \{S \in \mathcal{OBG} : E(S) \in \mathcal{U}\ ,\ H_e \in \mathcal{V}\ ,\ \text{for all}\ \ e \in E(S)\}. \qquad (2)$$

This result is a model for all attempts to describe sublattices of $L(\mathcal{CR})$ of the form $L(\mathcal{V})$ since it describes $L(\mathcal{OBG})$ as a <u>full</u> direct product of two better known (even completely known, in the case of $L(\mathcal{B})$) lattices. In general, one must be satisfied with a description as a subdirect product. Moreover, an element $\mathcal{W}$ in $L(\mathcal{OBG})$ is easily recovered from its image $\mathcal{W}\chi$ either lattice theoretically (as in (1)) or by identifying its members (as in (2)).

4. COMPLETELY SIMPLE SEMIGROUPS

The class $\mathcal{CS}$ of completely simple semigroups was the first class of completely regular semigroups, other than groups, for which a good representation theorem was obtained. Indeed, the Rees matrix representation of completely simple semigroups was perhaps so good that it appeared that any question relating to completely simple semigroups could readily be answered by means of a quick calculation in a Rees matrix semigroup. Perhaps this delayed interest in the lattice of varieties of completely simple semigroups, which has turned out to be anything but simple.

The class $\mathcal{CS}$ of completely simple semigroups constitutes a variety:

$$\mathcal{CS} = [(xyx)^0 = x^0]\ .$$

The lattice $L(\mathcal{CS})$ of subvarieties of $\mathcal{CS}$ has been studied in some depth in recent years: Jones [19], Masevickii [24], Petrich and Reilly [35], [36], [37], [38] and Rasin [41], [42].

However, even taking the structure of $L(\mathcal{G})$ for granted, only a relatively small part of $L(\mathcal{CS})$ is well understood. Much more so than in other varieties, the study of $L(\mathcal{CS})$ has relied heavily on the description of the free completely simple semigroup.

Theorem 4.1 (Clifford [5]). Let $X = \{x_i : i \in I\}$ be a non-empty set. Fix $1 \in I$ and let $I' = I \setminus \{1\}$. Let $Y = \{p_{jk} : j,k \in I'\}$, $Z = X \cup Y$, $G = G_Z$ be the free group on Z, $p_{1k} = p_{j1} = 1_G$ $(j,k \in I)$ and $P = (p_{jk})$ be the $I \times I$ matrix with $(j,k)^{th}$ entry equal to p_{jk}. Then the free completely simple semigroup on X is

$$F_X^{CS} = M(I,G,I;P)$$

together with the embedding $x_i \to (i,x_i,i)$.

Let $|X| = \aleph_0$ and $E(G_Z)$ denote the set of endomorphisms ω of G_Z such that for some $\theta,\varphi \in T_I$

$$p_{jk}\omega = p_{1\varphi 1\theta}\, p_{j\varphi 1\theta}^{-1}\, p_{j\varphi k\theta}\, p_{1\varphi k\theta}^{-1} .$$

The correspondence in the next result can be thought of as the counterpart of the anti-isomorphism between the lattice of group varieties and the lattice of fully invariant subgroups of the free group. We will denote by RB the variety of rectangular bands: $RB = [xyx = x]$.

Theorem 4.2 (Rasin [41]). The interval $[RB,CS]$ is a complete modular lattice anti-isomorphic to the lattice N of those normal subgroups of G_Z that are invariant under endomorphisms in $E(G_Z)$.

A coarse decomposition of $L(CS)$ is given by the following:

Theorem 4.3 (Petrich and Reilly [35]). The mapping

$$V \to (V \vee G\ ,\ V \cap G)$$

is an isomorphism of $L(CS)$ onto a subdirect product of the interval $[G,CS]$ and $L(G)$.

However, the structure of the lattice $[G,CS]$ is still largely obscure. Since its cardinality is $2^{\aleph_0}$ (Petrich and Reilly [36]), it seems unlikely that it could be completely described.

For any $S \in CR$, the core $C(S)$ of S is defined to be the subsemigroup $\langle E(S)\rangle$ generated by the idempotents $E(S)$ of S. For any variety $V \in L(CR)$, we define

$$VI = \langle C(S) : S \in V\rangle$$

the variety generated by $\{C(S) : S \in V\}$. For any group H, let $Z(H)$ denote the centre of H and let

$$C = \{S \in CS : C(S) \cap H_e \leq Z(H_e)\ ,\ \text{for all } e \in E(S)\}.$$

Then C is a variety. Indeed

$$\mathcal{C} = [ax^0a^0ya = aya^0x^0a].$$

The largest sublattice of $\mathcal{L}(\mathcal{CS})$ to have been given a reasonably complete description is $\mathcal{L}(\mathcal{C})$. Let $\mathcal{A}$ denote the variety of abelian groups: $\mathcal{A} = [x^0 = y^0, xy = yx]$.

<u>Theorem</u> 4.4 (Petrich and Reilly [37]). The mapping

$$\chi : \mathcal{V} \to (\mathcal{V} \cap \mathcal{RB}, \mathcal{VI} \cap \mathcal{A}, \mathcal{V} \cap \mathcal{G})$$

is an isomorphism of $\mathcal{L}(\mathcal{C})$ onto the subdirect product

$$\{(\mathcal{U},\mathcal{V},\mathcal{W}) \in \mathcal{L}(\mathcal{RB}) \times \mathcal{L}(\mathcal{AG}) \times \mathcal{L}(\mathcal{G}) : \mathcal{V} \subseteq \mathcal{W}, \text{ and } \mathcal{U} \neq \mathcal{RB} \Rightarrow \mathcal{V} = \mathcal{T}\}.$$

Moreover,

$$(\mathcal{U},\mathcal{V},\mathcal{W})\chi^{-1} = \{S \in \mathcal{CS} : S/\mathcal{H} \in \mathcal{U}, C(S) \cap H_e \in \mathcal{V}, H_e \in \mathcal{W}, \text{ for all } e \in E(S)\}.$$

Given this reasonably specific description of $\mathcal{L}(\mathcal{C})$, it is interesting to note $\mathcal{C}$ can, like $\mathcal{G}$ in Theorem 3.1, be used to give a representation of $\mathcal{L}(\mathcal{CS})$:

<u>Theorem</u> 4.5 (Petrich and Reilly [38]). The mapping

$$\mathcal{V} \to (\mathcal{V} \cap \mathcal{C}, \mathcal{V} \vee \mathcal{C})$$

is an isomorphism of $\mathcal{L}(\mathcal{CS})$ onto a subdirect product of $\mathcal{L}(\mathcal{C})$ and $[\mathcal{C},\mathcal{CS}]$.

However, the weakness in this result is that the interval $[\mathcal{C},\mathcal{CS}]$ remains mysterious. One variety that stands out in this interval is

$$\mathcal{D} = \{S \in \mathcal{CS} : C(S) \cap H_e \text{ is abelian, for all } e \in E(S)\}$$
$$= [ax^0a^0y^0a = ay^0a^0x^0a].$$

For even the interval $[\mathcal{C},\mathcal{D}]$ is extremely complex, containing, as it does, a set of $2^{\aleph_0}$ incomparable varieties.

5. COMPLETELY REGULAR SEMIGROUPS

In the attempt to describe ever larger ideals in $\mathcal{L}(\mathcal{CR})$, various results have been obtained which describe ideals of the form $\mathcal{L}(\mathcal{U})$ $(\mathcal{U} \in \mathcal{L}(\mathcal{CR}))$ as subdirect products of ideals $\mathcal{L}(\mathcal{V})$ and $\mathcal{L}(\mathcal{W})$ $(\mathcal{V},\mathcal{W} \in \mathcal{L}(\mathcal{CR}))$ where $\mathcal{V}$ and $\mathcal{W}$ are better understood than $\mathcal{U}$. Some results of this nature have been discussed in earlier sections. Hall and Jones [18] introduced the variety

$$\mathcal{LOBG} = \{S : eSe \in \mathcal{OBG} \text{ , for all } e \in E(S)\}$$

and showed that

$$\mathcal{LOBG} = \mathcal{CS} \vee \mathcal{B}$$

while Hall and Jones [18] and Rasin [43] independently, described $L(\mathcal{LOBG}) = L(\mathcal{CS} \vee \mathcal{B})$ as a subdirect product of $L(\mathcal{CS})$ and $L(\mathcal{B})$. In fact

Theorem 5.1 (Hall and Jones [18], Rasin [43]). The mapping

$$\mathcal{X} \to (\mathcal{X} \cap \mathcal{B} \text{ , } \mathcal{X} \cap \mathcal{CS})$$

is an isomorphism of the interval $[\mathcal{RB}, \mathcal{LOBG}]$ onto the full direct product $[\mathcal{RB},\mathcal{B}] \times [\mathcal{RB},\mathcal{CS}]$.

Going a step further we have:

Theorem 5.2 (Reilly [46]). The mapping

$$\chi : \mathcal{X} \to (\mathcal{X} \cap \mathcal{O}, \ \mathcal{X} \cap \mathcal{CS})$$

is an isomorphism of $[\mathcal{RB}, \mathcal{O} \vee \mathcal{CS}]$ onto the subdirect product of $L(\mathcal{O})$ and $L(\mathcal{CS})$

$$\{(\mathcal{U},\mathcal{V}) \in L(\mathcal{V}) \times L(\mathcal{CS}) : \mathcal{U} \cap \mathcal{CS} = \mathcal{V} \cap \mathcal{O}\} .$$

Moreover,

$$(\mathcal{U},\mathcal{V})\chi^{-1} = \{S \in \mathcal{CR} : S/\mu \in \mathcal{U}, \ eSe \in \mathcal{O} \text{ and } \mathcal{D}_e \in \mathcal{V}, \text{ for all } e \in E(S)\}.$$

The above result throws the description of $L(\mathcal{O} \vee \mathcal{CS})$ back to that of $L(\mathcal{O})$ and $L(\mathcal{CS})$. In [44], Rasin has described the lattice of varieties in $L(\mathcal{O})$ which have torsion subgroups while Gerhard and Petrich [14] have described the lattice of subvarieties of $\mathcal{ROG}$, this latter variety consisting of all those members S of $\mathcal{O}$ for which $E(S)$ is a regular band $(axya = axaya)$.

However, a very interesting recent development has given an indication of the complexity of $L(\mathcal{CR})$ and of how modest the achievements in describing ever larger ideals of $L(\mathcal{CR})$ have been to date relative to the full complexity of $L(\mathcal{CR})$. This leads us to the study of various operators on the lattice $L(\mathcal{CR})$ which, in a certain sense, span the lattice $L(\mathcal{CR})$. Such operators cannot be expected to reach $\mathcal{CR}$ itself in a finite numer of steps since we know:

Proposition 5.3 (Petrich and Reilly [39]). $\mathcal{CR}$ is finitely join irreducible.

As with any variety of algebras, there is an anti-isomorphism from $L(\mathcal{CR})$ to the lattice of fully invariant congruences on the free completely regular semigroup $F_X^{\mathcal{CR}}$ on a countably infinite set X. For any variety $\mathcal{V} \in \mathcal{L}(\mathcal{CR})$, let $\rho_{\mathcal{V}}$ denote the congruence on $F_X^{\mathcal{CR}}$ corresponding to $\mathcal{V}$ and for any fully invariant congruence ρ, let $[\rho]$ denote the variety defined by ρ.

<u>Notation</u>. For any congruence ρ on a completely regular semigroup S, let

$$\ker \rho = \{x \in S : x \,\rho\, x^0\} \qquad \operatorname{tr} \rho = \rho|_{E(S)}$$

$$\operatorname{ltr} \rho = (\rho \vee \mathcal{L})|_{E(S)} \qquad \operatorname{rtr} \rho = (\rho \vee \mathcal{R})|_{E(S)} .$$

Each of these objects determines an equivalence relation on the <u>lattice of congruences</u> $\operatorname{Con}(S)$ of S:

$$K = \{(\rho,\theta) : \ker \rho = \ker \theta\} \quad T = \{(\rho,\theta) : \operatorname{tr} \rho = \operatorname{tr} \theta\}$$

$$T_l = \{(\rho,\theta) : \operatorname{ltr} \rho = \operatorname{ltr} \theta\} \quad T_r = \{(\rho,\theta) : \operatorname{rtr} \rho = \operatorname{rtr} \theta\} .$$

The most fundamental observations on these relations are:
(i) (Feigenbaum [8]) $K \cap T = \iota$,
(ii) (Pastijn and Petrich [27]) $T_l \cap T_r = T$.

Moreover, for any congruence ρ the K, T, T_l and T_r classes all have minimum elements which we shall denote by

$$\rho_K, \ \rho_T, \ \rho_{T_l} \ \text{and} \ \rho_{T_r}$$

respectively. A key observation for the application of these ideas to $\mathcal{L}(\mathcal{CR})$ is the following:

<u>Lemma</u> 5.4 (Pastijn and Trotter [29] and Pastijn [26]). Let ρ be a fully invariant congruence on $F_X^{\mathcal{CR}}$. Then ρ_K, ρ_T, ρ_{T_l} and ρ_{T_r} are also fully invariant.

Thus, by passing from a fully invariant congruence ρ to ρ_K, ρ_T etc. we have a means of passing from a given fully invariant congruence ρ to related but smaller fully invariant congruences. What is more, this process can be repeated to yield $(\rho_K)_T = \rho_{KT}$, $(\rho_T)_K = \rho_{TK}$ etc. This leads to a network of fully invariant congruences on $F_X^{\mathcal{CR}}$ known as the MIN-NETWORK. Thus we have:

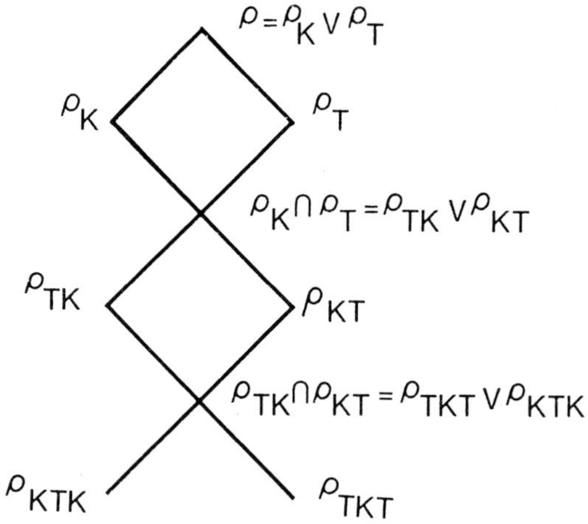

Figure 2

The correspondence between varieties and fully invariant congruences enables us to define four operators on $L(CR)$ as follows: for any $V \in L(CR)$, $\rho = \rho_V$ let $VK^* = [\rho_K]$, $VT^* = [\rho_T]$, $VT_l^* = [\rho_{T_l}]$, $VT_r^* = [\rho_{T_r}]$. Invoking the duality between fully invariant congruences and varieties, Figure 2 transforms into Figure 3 in terms of varieties.

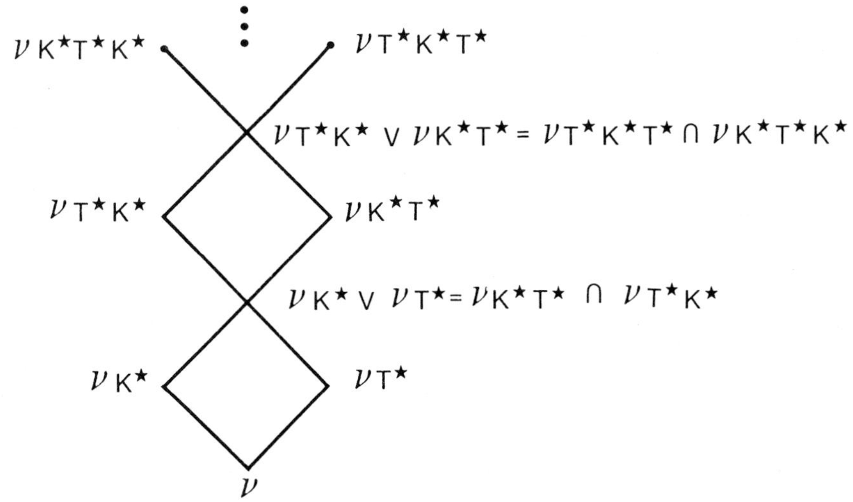

Figure 3

Theorem 5.5 (Pastijn and Trotter [29]). Let $V \in L(CR)$, $V \neq CR$. Then

$$V \subsetneq V(K^*T^*) \subsetneq V(K^*T^*)^2 \ldots \quad \text{and} \quad \vee V(K^*T^*)^n = CR .$$

Moreover, if A and B are incomparable varieties in the network, then

$$[A \cap B, A \vee B] \cong [A, A \vee B] \times [B, A \vee B] \cong [A \cap B, A] \times [A \cap B, B].$$

In order to work with T^*, K^*, T_l^* and T_r^* effectively, it is useful to have alternative descriptions.

Definition. Let $U, V \in L(CR)$. Then the Mal'cev product of U and V is

$$U \circ V = \{S \in CR : \exists\, \rho \in \mathrm{Con}(S) \text{ such that (i) } (x\rho)^2 = x\rho \Rightarrow x\rho \in U \text{ and (ii) } S/\rho \in V\} .$$

In general, the Mal'cev product is not a variety (see Jones [20]). However, if we let S denote the variety of semilattices, then

Theorem 5.6 (Jones [20]). (i) $U, V \in L(CS) \Rightarrow U \circ V \in L(CS)$.
(ii) $U \in L(RB \vee G)$, $S \subset V \Rightarrow U \circ V \in L(CR)$.

Recall that μ_S (respectively, τ_S) denotes the maximum idempotent separating (respectively, idempotent pure) congruence on a semigroup S. Also, for any equivalence relation λ on S, we denote by λ^0 the largest congruence on S contained in λ. We can now give two alternative descriptions of each of VK^*, VT^*, VT_l^* and VT_r^*.

Let LG (respectively, RG) denote the variety of left groups (respectively, right groups).

Theorem 5.7 (Jones [20], Pastijn and Trotter [29], Pastijn [26], Reilly [46]). Let $V \in L(CR)$. Then

(i) $VK^* = B \circ (V \vee S) = \{S \in CR : S/\tau_S \in V \vee S\}$,

(ii) $VT^* = G \circ V = \{S \in CR : S/\mu \in V\}$,

(iii) $VT_l^* = LG \circ V = \{S \in CR : S/L^0 \in V\}$,

(iv) $VT_r^* = RG \circ V = \{S \in CR : S/R^0 \in V\}$.

With the aid of these descriptions, we can now calculate some simple cases:

$$TT^* = G \qquad TK^* = B$$
$$RBT^* = CS \qquad GK^* = O$$
$$BT^* = BG \qquad CSK^* = LO$$

where $LO = \{S \in CR : eSe \in O, \text{ for all } e \in E(S)\}$ is the variety of locally orthodox completely regular semigroups. It is then particularly interesting to consider the network generated by applying K^* and T^*

to $\mathcal{T}$ (the trivial variety):

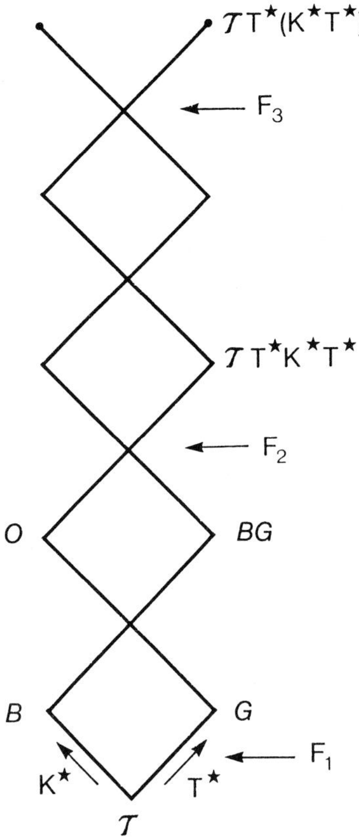

Figure 4

In Figure 4, F_n denotes the free completely regular semigroup on n generators and the arrows indicate the levels at which they are contained in the network.

Two other operations have played important roles in the study of $L(\mathcal{CR})$ to date.

Definition. For any variety $\mathcal{V} \in L(\mathcal{CR})$, let

$$\mathcal{V}_L = \{S : eSe \in \mathcal{V} , \text{ for all } e \in E(S)\}$$

$$\mathcal{V}_C = \{S : C(S) \in \mathcal{V}\} .$$

For example

$$\mathcal{T}_L = \mathcal{RB} \qquad \mathcal{G}_L = \mathcal{CS}$$
$$\mathcal{S}_L = \mathcal{NB} \qquad \mathcal{CS}_L = \mathcal{CS}$$

$$\mathcal{B}\mathrm{L} = \mathcal{B} \qquad \mathcal{BG}\mathrm{L} = \mathcal{BG} \qquad \text{(Hall and Jones [18])}$$
$$\mathcal{T}\mathrm{C} = \mathcal{G} \qquad \mathcal{B}\mathrm{C} = \mathcal{O}$$
$$\mathcal{RB}\mathrm{C} = \mathcal{RB} \vee \mathcal{G} \qquad \mathcal{BG}\mathrm{C} = \mathcal{G} \circ \mathcal{O}$$

Given the importance of the operators C,L,K,T in the study of $L(\mathcal{CR})$ it is natural to consider the semigroup of these operators qua semigroup. Their properties as operators are sufficiently well understood as to make this determination possible.

Theorem 5.8 (Petrich and Reilly [39]).
(i) The semigroup generated by the operators C,L is a four element semigroup {C,L,CL,LC} with a presentation in terms of generators and relations given by $\langle C,L : C^2 = C\ ,\ L^2 = L\ ,\ LCL = CLC = CL \rangle$.
(ii) The semigroup generated by the operators K^* and T^* has a presentation in terms of generators and relations given by $\langle K^*, T^* : (T^*)^2 = T^*, (K^*)^2 = K^* \rangle$.
(iii) The semigroup generated by the operators C,L,K^* and T^* has a presentation in terms of generators and relations given by

$$\langle C,L,K^*,T^* : C^2 = C,\ L^2 = L,\ (T^*)^2 = T^*, (K^*)^2 = K^*\ ,$$

$$CK^* = K^*C,\ CT^* = T^*C,\ LK^* = K^*L, LT^* = T^*L,\ LCL = CLC = CL\rangle\ .$$

(iv) In terms of monoids:

$$\langle C,L,K^*,T^*\rangle^1 \cong \langle C,L\rangle^1 \times \langle K^*,T^*\rangle^1\ .$$

The equivalence relations K,T,T_l,T_r defined earlier on the lattice of fully invariant congruences on $F_X^{\mathcal{CR}}$ determine (via the usual duality) equivalence relations on $L(\mathcal{CR})$ which we shall also denote by K,T,T_l, and T_r, respectively. These play a key role in the two most 'global' results on $L(\mathcal{CR})$ obtained thus far.

Theorem 5.9. The relations K,T,T_l and T_r are complete congruences on $L(\mathcal{CR})$. Moreover, the mapping

$$\mathcal{V} \to (\mathcal{V}T_l,\ \mathcal{V}K\ ,\ \mathcal{V}T_r)$$

is an isomorphism of $L(\mathcal{CR})$ onto a subdirect product of $L(\mathcal{CR})/T_l$, $L(\mathcal{CR})/K$ and $L(\mathcal{CR})/T_r$.

For the origins of the various parts of Theorem 5.9, see Pastijn [26].

Combining the fact that K is a congruence with a remarkable representation of $L(\mathcal{CR})$ as an ordered set of isotone mappings, due to Polák [40], Pastijn [26] established the following:

Theorem 5.10. $L(CR)$ is arguesian and therefore modular.

In conclusion, the recent work of Pastijn and Trotter on networks gives a certain perspective on all the results to date giving specific information or representations of sublattices of the form $L(U)$. All such varieties U are contained in $T(T^*K^*)^3$, while $CR = \bigvee_{n=1}^{\infty} T(T^*K^*)^n$.

REFERENCES

[1] A.P. Birjukov, 'Varieties of idempotent semigroups', Algebra i Logika, 9 (1970), 255-273.

[2] A.H. Clifford, 'Semigroups admitting relative inverses', Annals of Math. 42 (1941), 1037-1049.

[3] A.H. Clifford, 'The structure of orthodox unions of groups', Semigroup Forum, 3 (1972), 283-337.

[4] A.H. Clifford, 'A structure theorem for orthogroups', J. Pure Appl. Math. 8 (1976), 23-50.

[5] A.H. Clifford, 'The free completely regular semigroup on a set, J. Algebra 59 (1979), 434-451.

[6] A.H. Clifford and M. Petrich, 'Some classes of completely regular semigroups', J. Algebra 46 (1977), 462-480.

[7] P.H.H. Fantham, 'On the classification of a certain type of semigroup', Proc. London Math. Soc. (3) 10 (1960), 409-427.

[8] R. Feigenbaum, 'Regular semigroup congruences', Semigroup Forum 17 (1979), 373-377.

[9] C.F. Fennemore, 'All varieties of bands', Math Nachr. 48 (1971) I:237-252, II: 253-262.

[10] J.A. Gerhard, 'The lattice of equational classes of idempotent semigroups', J. Algebra, 15 (1970), 195-224.

[11] J.A. Gerhard, 'Subdirectly irreducible idempotent semigroups', Pacific Journal of Mathematics, 39 (1971) 669-676.

[12] J.A. Gerhard, 'Free completely regular semigroups I. Representation', J. Algebra 82 (1983), 135-142.

[13] J.A. Gerhard, 'Free completely regular semigroups II. Word Problem', J. Algebra 82 (1983), 143-156.

[14] J.A. Gerhard and M. Petrich, 'All varieties of regular orthogroups', Semigroup Forum 31 (1985), 311-351.

[15] J.A. Gerhard and M. Petrich, 'Varieties of bands revisited', Preprint.

[16] J.A. Gerhard and M. Petrich, 'Free bands and free *-bands', Glasgow Math. J.

[17] J.A. Green and D. Rees, 'On semigroups in which $x^r = x$', Proc. Cambridge Phil. Soc. 48 (1952), 35-40.

[18] T.E. Hall and P.R. Jones, 'On the lattice of varieties of bands of groups', Pac. J. Math. 91 (1980), 327-337.

[19] P.R. Jones, 'Completely simple semigroups: free products, free semigroups and varieties', Proc. Roy. Soc. Edinburgh Sect. A 88 (1981), 293-313.

[20] P.R. Jones, Mal'cev products of varieties of completely regular semigroups', preprint.

[21] J. Kaďourek and Libor Polák, On the word problem for free completely regular semigroups, preprint.

[22] L.G. Kovács and M.F. Newman, 'Hanna Neumann's problem on varieties of groups', Proceedings, Second International Conference on Group Theory, Canberra, 1973, Lecture Notes in Mathematics No. 372, 221-225.

[23] J. Leech, 'The structure of a band of groups', Mem. Amer. Math. Soc. 1 (1975), No. 157, 67-95.

[24] G.I. Maševickiĭ, 'On identities in varieties of completely simple semigroups over abelian groups', Contemporary Algebra, Leningrad (1978), 81-89 (Russian).

[25] H. Neumann, 'Varieties of groups', Springer, New York 1967.

[26] F. Pastijn, 'The lattice of completely regular semigroup varieties', preprint.

[27] F. Pastijn and M. Petrich, 'Congruences on regular semigroups', Trans. Amer. Math. Soc., 295 (1986), 607-633.

[28] F. Pastijn and M. Petrich, 'The congruence lattice of a regular semigroup', J. Pure Appl. Math.

[29] F. Pastijn and P.G. Trotter, 'Lattices of completely regular semigroup varieties', Pacific J. Math. 119 (1985), 191-214.

[30] M. Petrich, 'The structure of completely regular semigroups', Trans. Amer. Math. Soc. 189 (1974), 211-236.

[31] M. Petrich, 'Varieties of orthodox bands of groups', Pacific J. Math., 58 (1975), 209-217.

[32] M. Petrich, 'Certain varieties and quasivarieties of completely regular semigroups', Can. J. Math. XXIX (1977) 1171-1197.

[33] M. Petrich, 'On the varieties of completely regular semigroups', Semigroup Forum 25 (1982), 153-169.

[34] M. Petrich, 'A structure theorem for completely regular semigroups', preprint.

[35] M. Petrich and N.R. Reilly, 'Varieties of groups and completely simple semigroups', Bull. Austral. Math. Soc. 23 (1981), 339-359.

[36] M. Petrich and N.R. Reilly, 'Near varieties of idempotent generated completely simple semigroups', Algebra Universalis 16 (1973), 83-104.

[37] M. Petrich and N.R. Reilly, 'All varieties of central completely simple semigroups', Trans. Amer. Math. Soc. 280 (1983), 623-636.

[38] M. Petrich and N.R. Reilly, 'Certain homomorphisms of the lattice of varieties of completely simple semigroups', J. Austral. Math. Soc. (Series A) 37 (1984), 287-306.

[39] M. Petrich and N.R. Reilly, 'A semigroup of operators on the lattice of varieties of completely regular semigroups', preprint.

[40] L. Polák, 'On varieties of completely regular semigroups I', Semigroup Forum 32 (1985), 97-123.

[41] V.V. Rasin, 'On the lattice of varieties of completely simple semigroups', Semigroup Forum 17 (1979), 113-122.

[42] V.V. Rasin, 'Free completely simple semigroups', Mat. Zapiski Ural. Univ. 11 (1979), 140-151 (Russian).

[43] V.V. Rasin, 'On the varieties of Cliffordian semigroups', Semigroup Forum 23 (1981), 201-220.

[44] V.V. Rasin, 'Varieties of orthodox Clifford semigroups', Izv. Vyssh. Uchebn. Zaved. Mat., 1982, No. 11, 82-85 (Russian)

[45] D. Rees, 'On semi-groups', Proc. Cambridge Phil. Soc. 36 (1940), 387-400.

[46] N.R. Reilly, 'Varieties of completely regular semigroups', J. Austral. Math. Soc. (Series A) 38 (1985), 372-393.

[47] P.G. Trotter, 'Free completely regular semigroups', Glasgow Math. J.

[48] P.G. Trotter, 'Relatively free bands of groups', preprint.

[49] R.J. Warne, 'On the structure of semigroups which are unions of groups', Trans. Amer. Math. Soc. 186 (1973), 385-401.

[50] S. Wismath, 'The lattice of varieties and pseudovarieties of band monoids', Semigroup Forum, 33 (1986), 187-198.

[51] M. Yamada, 'Construction of inversive semigroups', Mem. Fac. Lit. Sci. Shimane Univ., Nat. Sci 4 (1971), 1-9.

NEW TECHNIQUES IN GLOBAL SEMIGROUP THEORY

John Rhodes
Department of Mathematics
University of California
Berkeley, California 94720

Herein I state new results of several researchers (including myself) and present some new conjectures. Almost no proofs are given, but instead, references and (hopefully) helpful remarks. The new techniques are to be learned by studying the references. This paper is just a signpost.

In the spirit of Paul Erdos, I offer U.S. dollar amounts for solutions of various conjectures and problems. Decisions of the judge (myself) are final regarding correctness, priority, exact statement, amount, etc.

(1.0) Notation (see [Ei]). Let A denote the pseudo-variety (PV) of finite *aperiodic* semigroups (i.e. finite semigroups whose subgroups are singletons).

Let G denote the PV of ***finite groups.*** Let **Com** denote the PV of ***finite commutative*** semigroups. Let **Nil** denote the PV of ***finite nilpotent*** semigroups. J and J_1 denote the pseudovarieties of ***finite*** $\mathcal{J}$***-trivial*** and ***finite idempotent and commutative*** semigroups. B denotes the pseudovariety of ***finite idempotent semigroups.***

If V_1, V_2 are PV's, then $V_1 \vee V_2 = \{S: S \prec S_1 \times S_2 \text{ for } S_j \in V_j\}$ is the *join* of V_1 and V_2. Here $\times$ denotes direct product and $\prec$ denotes division (i.e., "is a homomorphic image of a subsemigroup of") (see [Ei]).

(1.1) CONJECTURE ($25). Is membership of S in $A \vee G$ decidable? Equivalently, is there an algorithm that determines from the multiplication table of a finite semigroup whether S is a member of $A \vee G$?

(1.2) CONJECTURE ($10). Is membership of S in **Com** $\vee$ G decidable?

(1.3) PROBLEM ($50). Consider joins and intersections of the following PV's: A, G, **Nil**, J_1, J (e.g., G $\vee$ (**Nil** $\cap$ **Com**), A $\vee$ G $\vee$ **Com**, etc.). For which such is the membership problem undecidable?

S. M. Goberstein and P. M. Higgins (eds.), Semigroups and Their Applications, 169–181.

D. Albert and I [A+R] have shown that the PV $V_1 \vee V_2$ may have an ***undecidable*** memberhsip problem, even when PV's V_1 and V_2 have decidable membership problems (see below). Clearly all the PV's of Notation (1.0) have a decidable membership problem.

(1.4) Notation. Let E denote a ***finite*** set of equations in two variables. Let PV(E) denote the equational PV consisting of {S: S satisfies the equations of E}. By definition, PVM(E) = PV(E) ∩ Monoids. Since E is finite, membership in PV(E) and PVM(E) are clearly both decidable.

(1.5) Result (Albert and Rhodes [A+R]). There exist a finite set of equations E in two-variables such that

$$PV(E) \vee \mathbf{Com}$$

has an ***undecidable*** membership problem (clearly PV(E) and **Com** have decidable membership problems).

(1.6) Corollary of the proof of Result (1.5). The conclusion of (1.5) is unchanged when

$$PV(E) \vee \mathbf{Com}$$

is replaced by

$$PV(E) \vee (\mathbf{Com} \cap \mathbf{Nil})$$

or

$$PVM(E) \vee \mathbf{Com}$$

or

$$PVM(E) \vee (\mathbf{Com} \cap \mathbf{Nil}) \quad .$$

Key lemmas in the proof of (1.5) are the following:

(1.7) LEMMA (Albert and Rhodes, [A+R]). *There exists a finite set* E *of equations in two variables such that the following problem is **undecidable**. Given an equation* $\omega_1 = \omega_2$ *in two variables, determine whether every **finite** semigroup that satisfies* E *also satisfies the equation* $\omega_1 = \omega_2$.

Intuitively the result states that there exists a finite number of equations E which when we restrict to ***finite*** semigroups, implies some new "mysterious" infinite non-R.E. list of equations $\varepsilon(E)$ (whose complement is R.E.) true in all ***finite*** semigroups satisfying E. However, there exist infinite semigroups satisfying E which do not satisfy $\varepsilon(E)$. In fact, those members of $\varepsilon(E)$ which are not satisfied by some infinite semigroup form a non-R.E. subset (whose complement is R.E.).

(1.8) Corollary of the Proof of Lemma (1.7). In Lemma (1.7), $\omega_1 = \omega_2$ can be restricted to those $\omega_1 = \omega_2$ such that ω_1 and ω_2 have exactly the same number of occurrences of each variable, and the undecidable conclusion is still valid.

Further, we can also restrict to ***monoids.***

The proof of (1.8) and (1.9) requires "coding Turing machines" (see [A+R]). The other lemma required is the following.

(1.9) REDUCTION LEMMA [A+R]. *Membership in* PV(E) ∨ **Com** *is decidable implies the problem considered in Lemma (1.7) with the restriction of (1.8) is decidable.*

Since we know by (1.7) and (1.8) this last problem is not decidable, we have (1.5).

The proof of (1.9) is not difficult, but fundamental.

There exist several other important binary operations on PV's other than join, namely semi-direct or wreath product $*$ ([Ei],[Cat],[M]), the double semi-direct or block product $\square$ ([R+T],[M]), the diamond or Schützenberger product $\Diamond_A$ ([Ei],[B-R],[St]), and m, the Mal'cev product (defined next).

<u>(1.10) Definition.</u> Let V_j be a PV of finite semigroups. Then $V_2 m V_1$ read the ***Mal'cev product*** of V_2 and V_1, by definition equals $\{S$: there exists T such that $S \prec T$ and there exists surmorphism $\phi: T \twoheadrightarrow A_1 \in V_1$ such that $\forall\, e^2 = e \in A_1$, $\phi^{-1}(e) \in V_2\}$. For more details on m see [Exp], where the Mal'cev prodcut is first defined for PV's and the basic properties are developed.

We note the products join, $*$ are associative binary operators on PV's, $\Diamond$ is associative if we "cut down to generators" ([St],[B-R]) and $\square$ and m are ***not*** associative but

(1.11) PROPOSITION. For PV's V_j, $j=1,2,3$,

(1.11a) [R-T] $\qquad V_3 \square (V_2 \square V_1) \supseteq (V_3 \square V_2) \square V_1$,

(1.11b) [Exp] $\qquad V_3 m (V_2 m V_1) \subseteq (V_3 m V_2) m V_1$

Note the inclusions reverse (see [R-T] and [Exp]).

(1.12) CONJECTURE. ($50) Find PV's V_1, V_2 with ***decidable*** membership problems so that

(a) $V_2 * V_1$, (b) $V_2 \square V_1$, (c) $V_2 \Diamond V_1$, (d) $V_2 m V_1$

have ***undecidable*** membership problems. We know by (1.5) that this is true for join. (Note: Using the techniques of (1.5)-(1.9) plus the unpublished Presentation Lemma of Rhodes [PL], Albert and Rhodes [A+R] have a proposed proof of (a), (b) and (d), which is being checked and holding at the time of this writing.)

(1.13) GENERALIZED PROBLEM ($100 variously allotted). Consider well formed formulas of operations V, *, □, ◇, ∩ and PV's A, G, **Com**, **Nil**, J_1, J, B (e.g. $A * G * A$ or $((A*G) \vee G*A) \square J_1)$). Find a short, simple, "most natural" expression which has an ***undecidable*** membership problem.

We now turn to the Margolis conjecture, Ash's proof, immediate generalizations, and the important "Type-II Conjecture" of which these others are very special cases.

The Margolis (orthodox) conjecture is: the PV generated by the finite inverse (orthodox) semigroups equals those finite semigroups whose idempotents commute (form a subsemigroup) and hence form a semilattice (band). These problems are of the "must deal with the non-regular or null elements" type. Ash [Ash] brilliantly proved the Margolis conjecture, but once it is viewed as a special case of the "Type-II Conjecture" (see below), which is how Margolis viewed it when he proposed it, the orthodox conjecture follows also, and the proof is perhaps conceptually and technically easier. See [B-M-R] in this volume. The link between the Margolis/orthodox conjecture and the "Type-II Conjecture" is from Margolis and Pin [M-P] which is summarized in the following (1.14) and (1.18).

(1.14) Fact. (a) The PV generated by finite inverse semigroups is $J_1 * G$. See Notation (1.1).

(b) The PV generated by finite orthodox semigroups contains $B*G$.

Proof (Outline). For unitary semi-direct products, $R*G$ is regular, if R is regular and G is a group. Hence $S*G$ is inverse (orthodox) if $S \in J_1$ ($S \in B$). It's well known (the Tilson map) S inverse implies $S \prec (\{0,1\},\cdot) \circ G$, G a symmetric group.

We now precisely state the "Type-II conjecture", both strong and weak forms. See [K-R], [LBII], [B-R-M], [Ti-II] (the last two are in this volume).

(1.15) Definition (Rhodes and Tilson, [LBII]). Let S be a finite semigroup.

(a) $(S)_{II}$ the ***type-II subsemigroup*** of S, is defined by $s \in (S)_{II}$ iff for all finite groups G and for all subdirect products $R \leq\leq S \times G$, $(s,1) \in R$.

(b) $(S)_{II'}$, the ***type-II construct subsemigroup*** of S, is defined by

(i) $e^2 = e \in S \longrightarrow e \in (S)_{II'}$

(ii) $s_1, s_2 \in (S)_{II'} \longrightarrow s_1 \cdot s_2 \in (S)_{II'}$.

(iii) If $a,b \in S$, $aba = a$ (but not necessarily $bab = b$) and $s \in (S)_{II'}$, then $asb, bsa \in (S)_{II'}$.

(1.16) CONJECTURE ($100). The "Type-II conjecture" (see [K-R])

(a) (strong form): $(S)_{II} = (S)_{II'}$

(b) (weak form): $(S)_{II}$ is computable given finite S.

Clearly (a) → (b). $(S)_{II'}$ is clearly computable.

Rhodes and Tilson in [LBII] (see also [Ti-II] in this volume) proved

(1.17) Result. For all finite S,

(a) $(S)_{II'} \subseteq (S)_{II}$,

(b) For all regular $\mathcal{J}$-classes J of S (we do not assume S is regular), $J \cap (S)_{II'} = J \cap (S)_{II}$,

(c) $S \in A*G$ iff $(S)_{II'}$ is aperiodic iff $(S)_{II}$ is aperiodic, so membership in $A*G$ is decidable.

It is easy to verify if S is finite inverse or orthodox, then $(S)_{II'}$ is the set of idempotents of S. Further, Margolis-Pin in in [M-P] show (applying the canonical theology, see [M]),

(1.18) Result (Margolis-Pin [M-P]).

(a) $S \in J_1*G$ iff $(S)_{II} \in J_1$, so the Margolis conjecture is true iff (1.16)(a), the strong type-II conjecture, holds for finite inverse.

(b) $S \in B*G$ iff $(S)_{II} \in B$, so the orthodox conjecture is true if (1.16)(a), the strong type-II conjecture, holds for finite orthodox.

Proof (Rough outline). The proof uses the new category theory approach. See [Cat],[M],[R-T].

S to $(S)_{II}$ is functorial. See [LBII]. Almost from the definitions, given finite semigroup S, there exists a finite group G and a relational morphism $\Phi: S \twoheadrightarrow G$ such that the category C equal to D_Φ on K_Φ (see [Cat] or [R-T]) is such that for all objects c of C, the monoid $C(c,c)$ divides $(S)_{II}$, i.e., C is "locally" in $(S)_{II}$.

If $(S)_{II} \in J_1(B)$, then $C \prec M$ with $M \in J_1(B)$ by a well known theorem of Simon (Therien). Hence by the covering lemma (see [Cat] or [R-T]), $S \prec M*G$ and $M*G$ is inverse (orthodox).

Hence, using (1.18), to prove the Margolis (orthodox) conjecture, we must show that if $(S)_{II'}$ is a semilattice (band) then ***no element of*** $(S)_{II}$ ***is null***, i.e., $(S)_{II'} = (S)_{II}$, the strong "Type-II conjecture" in these special cases. The [B-M-R] form of Ash's proof does this.

(1.19) Result.

(a) [Ash], [B-M-R]. If $(S)_{II'} \in J_1(B)$, then $(S)_{II} = (S)_{II'}$. Hence the Margolis and orthodox conjectures are true.

(b) [B-M-R]. A necessary and sufficient condition that no ***null*** element of S is type II is that $(S)_{II'}$ is regular, which implies $(S)_{II} = (S)_{II'}$ in this case.

Hence under the hypothesis $(S)_{II'}$ is regular, the strong "Type-II conjecture" is true generalizing Ash's theorem. This is the strongest result to date. Also see [Exp].

An interesting class of infinite semigroups are those satisfying $(S)_{II}$ is completely regular (i.e., $(S)_{II}$ is a Clifford semigroup or is a union of groups). This condition is closely related to the condition that the union of the maximal subgroups form a subsemigroup. In fact, in the finite case these conditions are equivalent and equivalent with the condition that the idempotents generate a union of groups. Restricting to finite, (1.19)(b) implies $(S)_{II} = (S)_{II'}$. Let U denote the PV of finite union of groups semigroups. Then the proof of (1.18) can be extended to show $S \in U*G$ iff $(S)_{II'} \in U$ iff the subsemigroup generated by the idempotents of S is in U. To extend (1.18) we must use Therien's result that U is local.

We next consider which PV of finite semigroups are equational, see [Ei].

(1.20) Fact. (a) All PV's of finite semigroups are eventually equational, see [Ei].

(b) The PV V is equational iff $S_i \in V$ for $i=1,2,3,\ldots$ and $S < \prod_{i=1}^{\infty} S_i$ and S is ***finite*** implies $S \in V$. Further, the S_i can be restricted to "Birkhoff" generators and the iff still holds. Here "Birkhoff" generators of V means a subset $\{S_i\}$ of V such that all other members of V divide the infinite product $\prod_{i=1}^{\infty} S_i$. See [A+R].

By definition, a PV V is ***bounded-torsion*** iff there exists a positive integer n such that V satisfies $X^n = (X^n)^2 = X^{2n}$, i.e., n^{th} powers are idempotents. Clearly the PV of all finite semigroups is not equational and any V such that $V \supseteq \mathbf{Nil}$ is not equational. We wish to find some bounded-torsion PV's which are not equational. The following definitions are due to me.

(1.21) Definition. For each positive integer n, B_n is the PV = $\{S$: S is a finite semigroup satisfying $X^n = X^{2n}$ and the subgroups of S are commutative$\}$. By definition $ECB_n = B_n \cap PV(\text{finite-inverse}) = \{S$: $S \in B_n$ and idempotents of S

commute}. Finally, $PVIB_n = PV(B_n \cap \text{inverse})$. Note if the word "subgroups" in the definition is replaced by "group divisors", the two definitions are equivalent.

(1.21a) Result. For n *odd* and n large, B_n, ECB_n, $PVIB_n$ are *not* equational.

Proof (Very rough outline). We use the Adian-Novikov [A-N] result on the non-finiteness of the Burnside group for exponent n *odd* and $n \geq 665$, and [B-R/G] and the contra-positive of Fact (1.20)(b). For p a prime $p^2 \geq 665$, $Z_p \circ Z_p$ is 2-generated with exponent p^2. Hence $Z_p \circ Z_p$ is a homomorphic image of $B(2, X^{p^2} = 1) = B$, the 2 generated Burnside group and B is infinite and the finite subgroups are cyclic by [A-N]. Then $\hat{B}^2$ has ideal filtration (see [B-R/G]) $\hat{B}^2 = I_0 \supset I_1 \supset I_2 \supset \ldots \supset I_n \supset \ldots$ with $\cap I_n = \Phi$ with $\hat{B}^2/I_j \in PVIB_n$, $n \geq 3p^2$. But for all n, $Z_p \circ Z_p \notin B_n$, ECB_n, $PVIB_n$ since $Z_p \circ Z_p$ is a non-commutative finite group.

(1.22) CONJECTURE ($300). For all n sufficiently large, show that B_n, ECB_n, $PVIB_n$ are *not* equational.

(1.23) Remark. (a) Result (1.21a) seems rather surprising to me for the following reasons. Suppose V is a bounded-torsion PV satisfying $X^n = X^{2n}$. Then $(yz)(xyz)^n$ and $z(xyz)^n$ describe ***two arbitrary regular $\mathcal{L}$-equivalent elements of*** $S \in V$. Dually, $(xyz)^n x$, $(xyz)^n xy$ describe two ***arbitrary*** regular R-equivalent elements of $S \in V$. Similarly, $z(xyz)^n$, $(xyz)^n x$ describe ***two arbitrary regular $\mathcal{G}$-equivalent elements of*** $S \in V$. Hence $V = \{S: S \text{ satisfying } X^n = X^{2n}$ and regular $\mathcal{G}$-classes of S are commutative groups$\}$ is equational, namely, $V = PV(X^n = X^{2n}, \alpha = z(xyz)^n, \beta = (xyz)^n x, \alpha\beta = \beta\alpha)$. Similarly if regular $\mathcal{L}$-($\mathcal{R}$-)classes form a commutative group.

However, (1.21a) says we ***cannot*** write two arbitrary regular $\mathcal{H}$-equivalent elements in this form.

(1.23a) Problem.

(a)($50) Give a short or different proof of (1.21a) not using the Burnside problem results.

(b)($5) Is $ECB_n = PVIB_n$?

(c) Conjecture (1.22) is equivalent with: for $m = 2^k$, k large, B_n, ECB_n, $PVIB_n$ are ***not*** equational, using (1.21a).

CONJECTURE (1.24). ($500) For all positive integers a,b, $X^a = X^{a+b}$ has a decidable word problem? That is, for all positive integers a,b, any finitely generated free semigroup satisfying the identity $X^a = X^{a+b}$ has a decidable word problem.

A special case of Conjecture (1.24) is

CONJECTURE (1.25) ($200). For all positive integers n, $X^n = X^{n+1}$ has a decidable word problem?

Remarks and Problems (1.26). (a) $X^{a+1} = X$ has a decidable word problem for a *odd,* $a \geq 665$ by Adian [A-N].

(b) Is Free(k, $X^n = X^{n+1}$) residually finite for all (sufficiently large) n? Conjecture ($50) No! (Of course, residually finite implies decidable word problem.)

(c) By definition the fixed equations E satisfy (*) iff $E \longrightarrow \omega_1 = \omega_2$ (for $\omega_1 = \omega_2$ an arbitrary equation) in ***finite*** semigroups is undecidable.

Result (1.7) states such E exist.

CONJECTURE (1.26).

(a)($100) For what a,b does $X^a = X^{a+b}$ satisfy (*)?

(b)($50) For what n does $X^n = X^{n+1}$ satisfy (*)?

CONJECTURE (1.27) ($1000). Complexity c is ***undecidable*** (see [Ti] in [Ei], Sc = 0 iff $S \in A$, Sc=1 iff $S \in A*G*A$ and $S \notin A$, etc.) and 2-sided complexity C (see [R-T]) is ***undecidable***?

We know A*G, G*A have decidable membership problems. See [K-R], Is the membership problem for A*G*A undecidable? We believe Conjecture (1.27) is true!

We know the PV's $c_n = \{S: Sc \leq n\}$ for $n \geq 1$ and $C_n = \{S: SC \leq n\}$ for $n \geq 1$ are ***not*** local [Rhodes, unpubl.],[R-T].

Restricting to aperiodics A the question is

CONJECTURE (1.28) ($200). Is dot-depth decidable? For definitions see [Ei] and [St,dd].

We believe (1.28) could go either way!

Slightly weaker is the following "dot-depth + ε conjecture". For notation see [R-T] or [St,dd]. Let ℓ_1 denote the locally trivial finite categories, i.e. those finite categories C such that for all objects c, $C(c,c) = 1$. See [Cat]. Let the ***Straubing hierarchy*** of monoidal PV's be $S_0 = \{1\}$, $S_{n+1} = J \square \ell_1[S_n]$; let $T_0 = \{1\}$, $T_{n+1} = J_1 \square \ell_1[T_n]$; let $R_0 = \{1\}$, $R_{n+1} = \ell_1[J_1 \square \ell_1[R_n])$ $= \ell_1[J \square \ell_1(R_n)]$, the ***Tilson hierarchy.*** See [R-T]. Then for all $n \geq 0$, $\ell_1[T_n] = \ell_1[S_n] = \ell_1[R_n] = R_n$. Let D_n be the monoidal dot-depth hierarchy, see [St,dd]. It is well known $\cup S_i = \cup T_i = \cup R_i$ $= \cup D_i = A$ the aperiodics.

The strong form of (1.28) is (a version of the Straubing conjecture).

CONJECTURE (1.29) ($100). (The strong form of (1.28) or the "d-d+ε conjecture".) Is $\ell_1[T_n] = \ell_1[S_n] = \ell_1[R_n] = R_n$ equal to $\ell_1[D_n]$ for all n ?

Our last topic concerns Tilson's new ordering on infinite categories, on finite categories (FCAT), and finite directed multigroups (FDG). See [Cat] or [Rh, CDTG]. A $G \in$ FDG with n objects (or vertices) can be coded by an n×n matrix with coefficients in nonnegative integers, e.g.

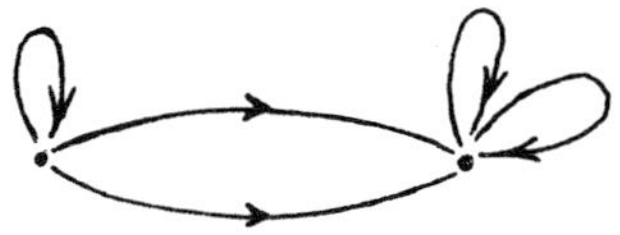

is coded by $\begin{pmatrix} 1 & 2 \\ 0 & 2 \end{pmatrix}$ (matrix of G) with

$$G(v_1) \equiv G(v_1,v_1) = 1$$

$$G(v_1,v_1) = 2$$

$$G(v_2,v_1) = 0$$

$$G(v_2) \equiv G(v_2,v_2) = 2$$

We consider finite categories (FCAT), finite semigroupoids (equals finite categories, perhaps with no identities) denoted FSgoid; finite directed multigroups FDG; finite transitive directed multigraphs (transitive means $G(v_1,v_2) \neq \emptyset$ and $G(v_2,v_3) \neq \emptyset$ implies $G(v_1,v_3) \neq \emptyset$) FTDG, finite graphs (symmetric DG's) denoted FG.

The objects and arrows of a category without the multiplication is a TDG. The possible support for FSgoid are exactly FTDG.

Minor Problem (1.30) ($5). What are the possible supports for FCAT's?

For $n \geq 0$ let $\vec{n} \in$ FDG have n+1 vertices (or objects) $\{0,1,\ldots,n\}$ and $\vec{n}(i,j) = 1$ iff $i < j$, so $\vec{2}$ equals

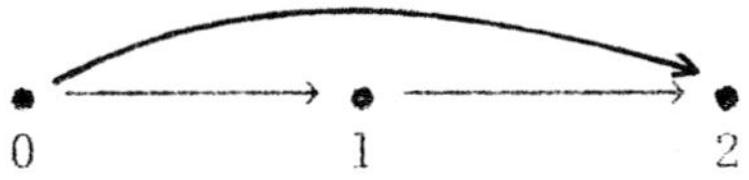

with matrix

$$\begin{pmatrix} 0 & 1 & 1 \\ 0 & 0 & 1 \\ 0 & 0 & 0 \end{pmatrix}$$

Now the important new definition from Tilson's [Cat] is

Definition (1.30)(a). Let $G_1, G_2 \in$ DG be two directed multigraphs. Then $G_1 \prec G_2$ read G_1 divides G_2 iff there exists a function f: vertices$(G_1) \to$ vertices(G_2) such that for all $v_1, v_1' \in$ vertices(G_2) such that for all $v_1, v_1' \in$ vertices(G_1)

$$|G_2(v_1 f, v_1' f)| \geq |G_1(v_1, v_1')|$$

where $|X|$ denotes cardinality of X. Thus for all n, $\vec{n} \prec (1)$ where (1) is the multigraph

$G_1 \precneqq G_2$ iff $G_1 \prec G_2$ and not $(G_2 \prec G_1)$. $G_1 \sim G_2$ iff $G_1 \prec G_2$ and $G_2 \prec G_1$. $G_1 <g\, G_2$ means G_1 is covered by G_2, i.e. $G_1 \precneqq G_2$ and $G_1 \prec G \prec G_2$ implies $G_1 \sim G$ or $G_2 \sim G$. $G_1 \prec g\, G_2$ is called a ***gap***. For references see Tilson's [Cat] or [Rh, CDTG].

Using the Dedekind height function it can be shown [Rh, CDTG] $\phi <g\, \vec{0} <g\, \vec{1} <g \ldots <g\, \vec{n} <g \ldots <g\, (1)$ are all the distinct divisors of (1) up to $\sim$ in FSgoid (not in FDG).

In the following let θ be any of FCAT, FSgoid, FDG, FTDG, FG. Using a "Ramsey-type" argument, one can show

Result (1.31) (Rhodes, [Rh, CDTG]). Let $G_1, G_2 \in \theta$. Then for all n,

$$G_2 \times \vec{n} \prec G_1 \text{ iff } G_2 \prec G_1$$

(Here $\times$ is direct product, i.e.,

$\text{vertices}(G \times H) = \text{vertices}(G) \times \text{vertices}(H)$ and

$$|G \times H((v_1, v_2), (v'_1, v'_2))| = |G(v_1, v'_1)| \cdot |G(v_2, v'_2)|) \quad .$$

COROLLARY (1.32) ([Rh, CDTG]). $G_1, G_2 \in \theta$, $G_1 \lneqq G_2$ implies there exist n such that $G_1 \lneqq G_1 \vee G_2 \times \vec{n} \leq G_2$. (Here $\vee$ is disjoint union.)

<u>Proof</u>. There exist n such that $G_2 \times \vec{n} \ngeq G_1$ if $G_1 \lneqq G_2$ by (1.31).

COROLLARY (1.33) ([Rh, CDTG]). $G_1 <g\ G_2$ and G_2 connected implies $G_2 \leq G_2 \times \vec{n}$.

<u>Proof</u>. Connected components $\leq$ connected components and (1.32).

<u>Definition (1.34)</u>. $G \in \theta$ is trivial iff there exists $n \geq 0$ such that $G \times \vec{n}$. (For FCAT adjoin identity arrows at each object.)

<u>Fact (1.35)</u> ([Rh, CDTG]).

(a) $C \in$ FCAT is trivial iff for all $c \in \text{Obj}(C)$, $C(c) = C(c,c) = 1$ the identity iff $C \in \ell_1$ (see [Cat]).

(b) $S \in$ FSgoid is trivial iff for all $c \in \text{Obj}(S)$, $S(c) = \emptyset$.

(c) $G \in$ FDG or FTDG is trivial iff it has ***no*** directed closed paths.

(d) $G \in$ FG or FTDG is trivial iff for all $v \in \text{vertices}(G)$, $G(v) = \Phi$, i.e., ***no*** loops.

<u>Result (1.36)</u> ([Rh, CDTG]). Let $\mathcal{E} = \{G \in \theta$: each connected component of G is non-trivial$\}$. Then $\mathcal{E}$ is ***dense***, i.e. $G_1, G_2 \in \mathcal{E}$, $G_1 \lneqq G_2$ implies there exists $G \in \mathcal{E}$ such that $G_1 \lneqq G \lneqq G_2$. Thus the ordering of the rational numbers is a suborder of $[G_1, G_2] \cap \mathcal{E}$ when $G_1 \lneqq G_2$.

In the above $\mathcal{E}$ can be replaced by the set of connected and non-trivial members of θ.

<u>General Problem (1.37)</u> (\$125). Find ***all*** the gaps in θ.

We break this problem down as follows.

CONJECTURE (1.37a) ($25). $\Phi <_g (1) < A_2$ (defined below) are the only gaps in connected finite categories (CFCAT)? More generally, find all gaps in CFCAT, FCAT.

Let $A_n = \begin{pmatrix} 1 & n \\ 0 & 1 \end{pmatrix}$. $A_n \in$ CFCAT. Then it is known that $A_2 \lneqq A_3 \lneqq A_4 \lneqq \cdots \lneqq A_n \lneqq \cdots$ are ***not*** gaps in CFCAT, see [Rh],[CDTG].

CONJECTURE (1.37 b-e). Classify all gaps in
(b)($25) FSgoid and CFSgoid.
(c)($25) FTDG and CFTDG.
(d)($25) FDG and CFDG.
(e)($25) FG and CFG.
(C means connected.)

Results (1.31)-(1.36) for θ show $G_1 <_g G_2$ and G_2 connected implies G_2 is trivial. Hence in θ "not too many gaps, many dense segments". Between Z_2 and Z_4 there exists a subset of θ order isomorphic to the rational numbers.

It is very interesting that dense sets (via Dedekind cuts) leads to real analysis, so writing down ***finite*** semigroup theory correctly (i.e. [Cat],[R-T]) immediately leads to dense orders between, say, Z_2 and Z_4 and hence real analysis.

REFERENCES

[A+R] D.Albert and J.Rhodes, "Undecidability of the identity problem for finite semigroups with applications" (preprint, 1986).

[Ash] C.Ash, "Finite semigroups with commuting idempotents," (preprint, 1986).

[A-N] S.I.Adian, <u>The Burnside Problem and Identities in Groups</u>, (Springer-Verlag, Berlin/New York, 1979) (Ergebnisse der Math. u. ihren Grenzgebiete 95), in Minicke (Ed.) Springer Lecture Notes in Math. No 806 (1980).)

[B-M-R] J.-C.Birget, S.Margolis, and J.Rhodes, "Finite semigroups whose idempotents commute or form a subsemigroup", in this volume.

[B-R] J.-C.Birget and J.Rhodes, "Almost finite expansions of arbitrary semigroups," Journal of Pure & Applied Algebra <u>32</u> (1984), 239-287.

[B-R/G] ———, "Group theory via global semigroup theory," Preprint from Center for Pure & Applied Mathematics, University of California, Berkeley, CA 94720, % J. Rhodes (December 1984).

[Cat] B.Tilson, "Categories as algebra: an essential ingredient in the theory of monoids," to appear in Journal of Pure & Applied Algebra.

[Ei] S.Eilenberg, Automata, Languages and Machines, vol. B, (Academic Press, New York, 1976).

[Exp] J.Rhodes, "Expansions of semigroups via Ramsey's and Brown's theorem," (preprint, 1986).

[K+R] J.Karnofsky and J.Rhodes, "Decidability of complexity one-half for finite semigroups," Semigroup Forum 24 (1984), 55-66.

[LBII] B.Tilson and J.Rhodes, "Improved lower bounds for the complexity of finite semigroups," Journal of Pure & Applied Algebra 2 (1972), 13-71.

[M] J.Rhodes, "Global structure theorems for arbitrary semigroups," in Proceedings of the 1984 Marquette Conference on Semigroups (K.Byleen, P.Jones, and F.Pastijn, eds.), pp.197-228, (Marquette Mathematics Dept., 1984).

[M-P] S.Margolis and J.-E.Pin, "I: Inverse semigroups and extensions of groups by semilattices; II: Expansions, for inverse semigroups and Schutzenberger products; III: Inverse semigroups and varieties of finite semigroups," to appear in Journal of Algebra.

[PL] J.Rhodes, "The presentation lemma for finite semigroups dividing A*G*V," (preprint, 1975, from Center for Pure & Applied Mathematics, University of California, Berkeley, CA 94720, % J.Rhodes).

[Rh, CDTG] J.Rhodes, "On the Cantor-Dedekind property of the Tilson order on finite categories and graphs," (preprint, 1984, from Center for Pure & Applied Mathematics, University of California, Berkeley, CA 94720, % J.Rhodes).

[R-T] J.Rhodes and B.Tilson, "The kernel of monoid morphism: a reversal invariant decomposition theory," (preprint, 1986, from Center for Pure & Applied Mathematics, University of California, Berkeley, CA 94720, % J.Rhodes).

[St] H.Straubing, "A generalization of the Schutzenberger product of finite monoids," Theoret. Comput. Sci. 13 (1981), 137-150.

[St,dd] ——— , "Monoids of dot-depth two," (preprint 1986).

[Ti] B.Tilson, Chapter XII in [Ei].

[Ti-II] ——— , "Type II redux," this volume.

FINITE J-TRIVIAL MONOIDS AND PARTIALLY ORDERED MONOIDS

Howard Straubing
Department of Computer Science
Boston College
Chestnut Hill, Massachusetts, USA 02167

Denis Thérien
School of Computer Science
McGill University
Montréal, Québec, Canada H3A2K6

ABSTRACT. An important result in the theory of automata, due to Imre Simon, characterizes the recognizable languages whose syntactic monoids are J-trivial. This theorem has an easy, although not entirely obvious, restatement as a global structure theorem for finite J-trivial monoids, asserting that every finite J-trivial monoid is a quotient of a finite monoid admitting a partial order compatible with the multiplication and having 1 as the maximum element. We have proved the theorem in this form using semigroup expansion techniques. In the present note we discuss the background and significance of this work and give a brief sketch of our proof. The full details of the proof will appear elsewhere [ST].

1. SIMON'S THEOREM

Let A be a finite alphabet. A^* denotes the free monoid generated by A, and A^+ the free semigroup generated by A. If $w \in A^*$ then $|w|$ denotes the length of w. The identity of A^* (the empty word) is denoted 1. Observe $|1| = 0$ and $A^+ = A^* \setminus \{1\}$.

If $v, w \in A^*$ we say v is a subword of w, and write $v \subset w$, if v is a

S. M. Goberstein and P. M. Higgins (eds.), Semigroups and Their Applications, 183–189.

subsequence of w--that is, v can be obtained from w by erasing some letters. In particular, $1 \subset w$ for all $w \in A^*$. We define

$$Sub_k(w) = \{v \in A^* \mid v \subset w;\ |v| \leq k\}$$

and

$$u \equiv_k v$$

if and only if $Sub_k(u) = Sub_k(v)$. (For example, $Sub_2(a^2b) = \{1,a,b,a^2,ab\}$ and $a^2b \equiv_2 a^rb$ for all $r \geq 2$.)

It is easy to show that each of the equivalence relations $\equiv_k$ is a finite congruence on A^*, and that the quotient monoids $A^*/\equiv_k$ are J-trivial--that is, they have one-element J-classes. The next theorem asserts that, up to homomorphic image, all finite J-trivial monoids arise in this fashion.

Theorem A. (I. Simon [Si1],[Si2]). Let M be a finite J-trivial monoid and $\varphi: A^* \longrightarrow M$ a morphism. Then φ factors through the projection morphism $\pi: A^* \longrightarrow A^*/\equiv_k$ for some $k \geq 0$.

Theorem A plays an important role in the algebraic theory of finite automata and the languages they recognize. It implies that a recognizable language L is a union of $\equiv_k$-classes for some k (L is piecewise testable) if and only if the syntactic monoid of L is J-trivial, and thus provides an effective criterion for identifying such languages. The family of piecewise testable languages forms the first level of a hierarchy of language classes based on the concatenation product (one version of the Brzozowski dot-depth hierarchy), and J-trivial monoids appear to be involved in the passage from each level of this hierarchy to the next. (See Thérien [The] and Straubing [V*D], [DD2]; also Thomas [Tho1], [Tho2] for a connection with formal logic and Barrington-Thérien [NC1] for a connection with computational complexity.)

A number of proofs of Theorem A have appeared in the literature (see Eilenberg [E] and Lallement [L] in addition to the references to Simon cited above). All of these proofs make use of nontrivial combinatorial properties of the congruences $\equiv_k$.

2. FINITE MONOIDS ADMITTING A PARTIAL ORDER

We wish to take a different view of Theorem A. Let us say that a monoid M is an integral p.o. monoid if M admits a partial order $\leq$ such that

(i) 1 is the maximum element of M with respect to $\leq$.

(ii) if $x \leq x'$ and $y \leq y'$ then $xy \leq x'y'$.

It is easy to see that the monoids $A^*/\equiv_k$ are p.o. monoids: We define $v/\equiv_k \leq w/\equiv_k$ if and only if $\mathrm{Sub}_k(w) \subseteq \mathrm{Sub}_k(v)$. We thus obtain

Theorem B. Every finite J-trivial monoid is a quotient of some finite integral p.o. monoid.

(Observe that every finite p.o. monoid, and hence every quotient of a finite integral p.o. monoid, is J-trivial, so the converse of Theorem B is true.)

In fact, we can also obtain Theorem A from Theorem B: Theorem B implies that to prove Theorem A, we need only consider the case where M is a finite integral p.o. monoid. Let k be the length of the longest strict chain with respect to the partial order on M, and let $w_1 \equiv_k w_2$. Then there is a factorization

$$w_1 = v_0 a_1 v_1 \ldots a_r v_r$$

where each $a_i \in A$, each $v_i \in A^*$, and $1 = v_0\varphi > (v_0 a_1)\varphi = (v_0 a_1 v_1)\varphi > (v_0 a_1 v_1 a_2)\varphi = \ldots$

Thus $r \leq k$ and $w_1\varphi = (a_1 \ldots a_r)\varphi \geq w_2\varphi$ (since $a_1 \ldots a_r$ is a subword of w_2).

Symmetrically $w_1\varphi \leq w_2\varphi$. Thus $w_1\varphi = w_2\varphi$, so φ factors through the projection morphism onto $A^*/\equiv_k$.

Theorem B is a global structure theorem for finite J-trivial monoids. It asserts that every finite J-trivial monoid M is the quotient of a finite monoid M' where M' is "close to M"--in this case J-triviality is preserved by the passage from M to M'--and where M' "admits a coordinatization such that multiplication is easy to perform in terms of the coordinates"--the partial order on M' can be viewed as a sort of coordinatization in this sense. More explicit coordinatizations, easily equivalent to Theorem B, are given in Straubing [J]. (The use of the term "global structure theorem" in this sense is apparently due to Rhodes. See [BR], which is the source of the quoted phrases above.)

We will prove Theorem B by giving an explicit construction of a finite p.o. monoid that maps onto a given finite J-trivial monoid M. The congruences $\equiv_k$ play no role in the proof. Instead we will use a semigroup 'expansion'. Semigroup expansions are studied in general in [BR]. It is remarkable that the same expansions, or slight variations thereof (e.g., the Rhodes expansion, the machine expansion, and, in our case, the Schutzenberger product) arise in a large number of different applications.

3. A SKETCH OF THE PROOF OF THEOREM B.

Let M be a finite J-trivial monoid. We proceed by induction on the cardinality of M. There is nothing to prove if M has one or two elements. Using the observation that the direct product of integral p.o. monoids is an integral p.o. monoid we may redcuce to the case where M has a unique 0-minimal ideal I, and every J-trivial monoid whose cardinality is smaller than that of M is the quotient of a finite integral p.o. monoid. Since M is J-trivial, $I = \{0, x\}$ for some x in M. If $M' = M\setminus\{0\}$ is closed under multiplication, then M' is the quotient of a finite integral p.o. monoid K, and we obtain a finite integral p.o. monoid that maps onto M by adjoining a new zero to K. If M' is not closed under multiplication, then it is not hard to show that $x^2 = 0$ (i.e., $\{x\}$ is a null J-class). The induction hypothesis implies that M/I is a quotient of a finite integral p.o. monoid K.

Our first impulse is to try to add some elements to the bottom of K to get an integral p.o. monoid that covers M--this is what we did in the case where M' was a submonoid of M. This does not appear to be possible: The manner in which the product of two elements of M-I falls to the ideal I may be quite erratic, and we cannot expect to recover this information from K. Instead we shall <u>expand</u> K as follows: Let ψ be the morphism from K onto M/I and η the usual morphism from M onto M/I. Let A be a finite alphabet and let $\varphi: A^+ \longrightarrow M$ be a surjective morphism such that $u\varphi = 1$ for some letter $u \in A$. There is thus a morphism $\alpha: A^* \longrightarrow K$ such that $\alpha\psi = \varphi\eta$ and $u\alpha = 1\alpha = .$ For $w \in A^+$ we define

$$X_w = \{(v\alpha, a, v'\alpha) | a \in A;\ v, v' \in A^*; w = vav'\}.$$

It is easy to verify that setting $X_v X_w = X_{vw}$ gives a well-defined associative product, so $K_1 = (X_w \mid w \in A^+)$ is a semigroup. (K_1 is a slight variant on the <u>Schutzenberger product</u> $K \diamond K$.) Of course K_1 maps onto K, and hence to M/I, but it is not difficult to show that $X_v = X_w$ implies $v\varphi = w\varphi$, so K_1 maps onto M as well. Let us define $X_v \ll X_w$ if for every triple $(k_1, a, k_2) \in X_v$ there is a triple $(k'_1, a', k'_2) \in X_w$ such that $k_1 \leq k'_1$ and $k_2 \leq k'_2$ in the partial order on K, and $a' = a$ or $a' = u$. One can then show that $X_v \ll X_w$ and $X_{v'} \ll X_{w'}$ implies $X_{vv'} \ll X_{ww'}$, and, with more difficulty, that $X_v \ll X_w$ and $X_w \ll X_v$ implies $v\varphi = w\,\varphi$.

The result is that the equivalence relation $\approx$ on K_1, defined by setting $X \approx Y$ if and only if $X \ll Y$ and $Y \ll X$, is a congruence, that the quotient $T = K_1/\approx$ is a semigroup admitting a partial order compatible with the multiplication, and that T maps onto M. $X_u T X_u$ is then a finite integral p.o. monoid that maps onto M.

4. REFERENCES

[BR] J.C. Birget and J. Rhodes, 'Almost finite expansions of arbitrary semigroups', *J. Pure and Applied Algebra* **32** (1984), 239-287.

[DD2] H. Straubing, 'Semigroups and languages of dot-depth 2', to appear in *Proc. 13th ICALP*, Springer Lecture Notes in Computer Science.

[E] S. Eilenberg, *Automata, Languages and Machines*, vol. B, Academic Press, New York (1976).

[J] H. Straubing, 'On finite J-trivial monoids', *Semigroup Forum* **19** (1980), 107-110.

[L] G. Lallement, *Semigroups and Combinatorial Applications*, Wiley, New York (1979).

[NC1] D. Barrington and D. Thérien, 'Finite monoids and the fine structure of NC1', preprint.

[Si1] I. Simon, Hierarchies of Events of Dot-Depth One', Ph. D. Dissertation, University of Waterloo (1972).

[Si2] I. Simon, 'Piecewise testable events' in Proc. 2nd GI Conference, Springer Lecture Notes in Computer Science **33** (1975), 214-222.

[ST] H. Straubing and D. Thérien, 'Partially ordered finite monoids and a theorem of I. Simon', submitted to *Journal of Algebra*.

[The] D. Thérien, 'Classification of finite monoids, the language approach', *Theoretical Computer Science* **14** (1981), 195-208.

[Tho1] W. Thomas, 'Classifying regular events in symbolic logic', *J. Computer and Systems Sciences* **25**, (1982),360-376.

[Tho2] W. Thomas, 'An application of the Ehrenfeucht-Fraissé game in formal language theory', Soc. mathématique de France, 2$^{\underline{e}}$ série, mémoire n$^{\underline{o}}$ 16 (1984).

[V*D] H. Straubing, 'Finite semigroup varieties of the form V*D', *J. Pure and Applied Algebra* **36** (1985), 53-94.

ON THE RECENT RESULTS IN THE STUDY OF POWER SEMIGROUPS

Takayuki Tamura
Department of Mathematics
University of California
Davis, California 95616
U.S.A.

ABSTRACT. This paper reports, without proof, that if S_1 and S_2 are finite semigroups and if their power semigroups are isomorphic, then S_1 and S_2 are isomorphic; and also reports some results on the study of $\mathcal{J}$-classes of the power semigroups of finite 0-simple semigroups. The detailed proofs will be published elsewhere [21], [22], [23].

1. INTRODUCTION

Let S be a semigroup and $\mathcal{P}(S)$ the power semigroup of S, i.e. the semigroup of all nonempty subsets of S in which the operation is defined by

$$XY = \{xy : x \in X,\ y \in Y\} \quad \text{for all} \quad X, Y \in \mathcal{P}(S).$$

Let $\mathcal{C}$ be a class of semigroups. In the study of the relationship between S and $\mathcal{P}(S)$ it is natural to consider the following three topics:

Topic I. Let $S_1, S_2 \in \mathcal{C}$. Does $\mathcal{P}(S_1) \cong \mathcal{P}(S_2)$ imply $S_1 \cong S_2$?

Topic II. Given a class $\mathcal{C}$ of semigroups, study the structure of $\mathcal{P}(S)$ for $S \in \mathcal{C}$.

Topic III. Given conditions on $\mathcal{P}(S)$, determine S for which $\mathcal{P}(S)$ satisfies them.

If the question in Topic I has an affirmative answer, $\mathcal{C}$ is said to be globally determined. Among the three topics, Topic I has been most actively studied. It is known that the following classes are globally determined: groups [15]; finite simple semigroups [2]; semilattices of torsion groups in which semilattices are finite [2]; rectangular groups [18]; semilattices [3]; completely 0-simple semigroups [19]. However the class of all semigroups is not globally determined [6], [7].

S. M. Goberstein and P. M. Higgins (eds.), Semigroups and Their Applications, 191–200.

Speaking of Topics II and III, the investigations proceeded in the following directions: the study of the power semigroup of the group of integers [11], [12], [13], [16], [20]; the semilattice decomposition of $\mathcal{P}(S)$ [9]; subgroups of $\mathcal{P}(S)$ [5], [10]; the study of finite $\mathcal{P}(S)$ related to language theory [4]; chain $\mathcal{P}(S)$ [17] and so on.

The three topics should be, of course, related to each other. For example, if the problem in Topic II is completely solved for a class $\mathcal{C}$, then the question in Topic I would be answered for $\mathcal{C}$, but Topic II is not necessarily a prerequisite for Topic I. For some $\mathcal{C}$ the global determination problem could be solved by using some properties of $S \in \mathcal{C}$ even if the structure of $S \in \mathcal{C}$ is not completely determined. Actually most cases of Topic I which have been considered follow this fashion. Nevertheless, the study of Topic II is interesting and has a good prospect.

In this paper we will treat Topics I and II. In particular, we will report that the class of finite semigroups is globally determined. Because of a limited number of pages, we will describe only the outline of the proof. We will also report a result on the study of $\mathcal{J}$-classes of $\mathcal{P}(S)$ of a finite 0-simple semigroup S.

As an illustration of how Topics II and III can interact, let us mention that $\mathcal{P}(S)$ contains zero if and only if S contains a group ideal. Some properties are possessed by both S and $\mathcal{P}(S)$: for example, S is a null semigroup or a left [right] zero semigroup if and only if $\mathcal{P}(S)$ is such; or S is an inflation of a rectangular band if and only if $\mathcal{P}(S)$ is an inflation of a rectangular band [18]. In this paper a similar result is given in Proposition 2.2. Topic III is expected to be more explored in the future. For undefined terminology the reader is referred to [1], [8].

2. GLOBAL DETERMINATION OF FINITE SEMIGROUPS

Let $\mathcal{C}$ be the class of finite semigroups. The <u>global determination</u> is strictly described as follows:

<u>If f is an isomorphism of $\mathcal{P}(S_1)$ to $\mathcal{P}(S_2)$, there is an isomorphism f' of $\mathcal{P}(S_1)$ to $\mathcal{P}(S_2)$ such that f' induces an isomorphism g of S_1 to S_2, that is,</u>

$$f' \epsilon_1 = \epsilon_2 g$$

<u>where ϵ_i is an embedding of S_i into $\mathcal{P}(S_i)$</u> $(i = 1,2)$.

THEOREM. <u>The class $\mathcal{C}$ of finite semigroups is globally determined</u>.

The proof is done by induction on the order, and the following obvious fact is used.

FACT 2.1. Let E_i and D_i be ideals of S_i such that $D_i \subset E_i$ $(i=1,2)$. Assume there is an isomorphism f of $\mathcal{P}(S_1)$ to $\mathcal{P}(S_2)$ such that f induces an isomorphism of D_1 to D_2 and an isomorphism of T_1 to T_2 where $T_i = E_i/D_i (i = 1,2)$. Then $E_1 \cong E_2$.

In the above, an isomporphism of T_1 to T_2 is interpreted as follows: the elements of $E_1 \backslash D_1$ are mapped to the elements of $E_2 \backslash D_2$ in a one-to-one and onto fashion and $x, y \in E_1 \backslash D_1$ and $xy \in D_1$ implies $f(xy) \in D_2$ and $f(x)f(y) \in D_2$.

A semigroup S is called nil if S contains zero 0 and if, for every $x \in S$, there is a positive integer n such that $x^n = 0$. An element a of S is called an annihilator of S if $ax = xa = 0$ for all $x \in S$, and the set of all annihilators of S is called the annihilator of S; it is denoted by $\mathcal{A}(S)$. If S is a finite nil semigroup, $\mathcal{A}(S) \neq \{0\}$ [14].

PROPOSITION 2.2. $\mathcal{P}(S)$ is finite nil if and only if S is finite nil. Furthermore

(2.2.1) $\quad \mathcal{A}(\mathcal{P}(S)) = \mathcal{P}(\mathcal{A}(S))$.

Accordingly, we divide $\mathcal{C}$ into the two subclasses: the class $\mathcal{C}_1$ of finite nil semigroups and the class $\mathcal{C}_2$ of finite non-nil semigroups.

Assume S_1 and S_2 are finite nil semigroups and let f be an isomorphism of $\mathcal{P}(S_1)$ to $\mathcal{P}(S_2)$. Then $\mathcal{A}(\mathcal{P}(S_1)) \cong \mathcal{A}(\mathcal{P}(S_2))$ which implies $\mathcal{A}(S_1) \cong \mathcal{A}(S_2)$ by induction hypothesis. Let $\mathcal{F}_1 = \mathcal{P}(S_1)/_{\mathcal{A}(\mathcal{P}(S_1))}$ and $\mathcal{F}_2 = \mathcal{P}(S_2)/_{\mathcal{A}(\mathcal{P}(S_2))}$. Then $\mathcal{F}_1 \cong \mathcal{F}_2$, so $\mathcal{A}(\mathcal{F}_1) \cong \mathcal{A}(\mathcal{F}_2)$. There exist $\mathcal{E}_i$ and E_i $(i = 1,2)$ such that

$$\mathcal{E}_{i}/_{\mathcal{A}(\mathcal{P}(S_i))} = \mathcal{A}\left(\mathcal{P}(S_i)\Big/_{\mathcal{A}(\mathcal{P}(S_i))}\right), \quad E_{i}/_{\mathcal{A}(S_i)} = \mathcal{A}\left(S_i/\mathcal{A}(S_i)\right), \quad i = 1,2.$$

It follows that $\mathcal{E}_1 \cong \mathcal{E}_2$ and $E_1 \cong E_2$. By repeating this process we can finally show that $S_1 \cong S_2$.

In case of non-nil semigroups, the situation is complicated. Let S be a finite semigroup with 0. There exists the largest nil ideal of S; it is called the nil-cover and denoted by $\mathcal{N}(S)$.

PROPOSITION 2.3. $\mathcal{P}(\mathcal{N}(S)) = \mathcal{N}(\mathcal{P}(S))$.

Accordingly if $\mathcal{P}(S_1) \cong \mathcal{P}(S_2)$, then $\mathcal{N}(\mathcal{P}(S_1)) = \mathcal{N}(\mathcal{P}(S_2))$. Other important parts are (0-) minimal ideal and the union of 0-minimal ideals. The regular socle of S is defined to be the union of all 0-simple ideals of S if S has 0 or the least ideal of S if S has no zero. The regular socle of S is denoted by $\mathcal{RS}(S)$. Now we have to close up the relation between a 0-simple ideal of S and a 0-simple ideal of $\mathcal{P}(S)$.

Let $D = \mathcal{M}^0(I,M,G;P)$ be the Rees regular representation of a completely 0-simple semigroup, i.e.

$$P = (p_{ji}),\ p_{ji} \in G^0 = G \cup \{0\}.$$

$$D = \{(i,j;x) : i \in I,\ j \in M,\ x \in G\} \cup \{0\},$$

$$(i,j;x)\ (k,\ell:y) = \begin{cases} (i,\ell;\ xp_{jk}y) & \text{if } p_{jk} \neq 0 \\ 0 & \text{if } p_{jk} \neq 0 \end{cases}$$

Define

$$\mathcal{I}_D = \{(A \times B;\ G)^0 : A \in \mathcal{P}(I),\ B \in \mathcal{P}(M)\} \cup \{0\}$$

where $(A \times B:G)^0 = (A \times B;G) \cup \{0\} = \{(i,j;x) : i \in A,\ j \in B,\ x \in G\} \cup \{0\}$,

$$(A \times B;G)^0 (C \times D;G)^0 = \begin{cases} (A \times D;G)^0 & \text{if } B\Delta_p C \\ 0 & \text{otherwise.} \end{cases}$$

$B\Delta_p C$ means that $p_{ji} \neq 0$ for some $j \in B,\ i \in C$.

LEMMA 2.4 [19] <u>$\mathcal{I}_D$ is a unique 0-minimal ideal of $\mathcal{P}(D)$.</u>

$\mathcal{I}_D$ is, of course, 0-simple. If $S = \mathcal{M}(I,M,G;P)$, then $\mathcal{I}_D = \{(A \times B;G) : A \in \mathcal{P}(I),\ B \in \mathcal{P}(M)\}$ is the least ideal of $\mathcal{P}(D)$.

On the other hand the following is concerned with maximal ideal.

LEMMA 2.5. <u>$\mathcal{P}(D)$ contains a unique maximal ideal $\mathcal{S}_D$ having the property that $\mathcal{R}_D = \mathcal{P}(D)\backslash\mathcal{S}_D$ contains at least an idempotent.</u>

From $\mathcal{R}_D$ and $\mathcal{I}_D$ we have

PROPOSITION 2.6. <u>Let D and D' be 0-simple ideals of S and S' respectively. If f is an isomporphism of $\mathcal{P}(S)$ to $\mathcal{P}(S')$ and if f induces $\mathcal{I}_D \cong \mathcal{I}_{D'}$ and $\mathcal{R}_D^0 \cong \mathcal{R}_{D'}^0$, then $D \cong D'$.</u>

The following gives a relation between 0-simple ideals of S and those of $\mathcal{P}(S)$.

PROPOSITION 2.7. <u>Let</u> D <u>be</u> <u>a</u> 0-<u>simple</u> <u>ideal</u> <u>of</u> S. <u>Then</u> $\mathcal{I}_D$ <u>is</u> <u>a</u> 0-<u>simple</u> <u>ideal</u> <u>of</u> $\mathcal{P}(S)$. <u>Moreover</u> <u>every</u> 0-<u>simple</u> <u>ideal</u> <u>of</u> $\mathcal{P}(S)$ <u>equals</u> $\mathcal{I}_D$ <u>for</u> <u>some</u> D <u>such</u> <u>as</u> <u>above</u>.

Let $\mathfrak{S}_D$ be the family of all ideals $\mathcal{I}$ of $\mathcal{P}(S)$ such that each $\mathcal{I}$ contains $\mathcal{I}_D$ as a unique 0-minimal ideal. For example, $\mathcal{I}_D$, $\mathcal{P}(D)$ and $\mathcal{S}_D$ are members of $\mathfrak{S}_D$. An isomorphism f of $\mathcal{P}(S)$ to $\mathcal{P}(S')$ induces $\mathfrak{S}_D \cong \mathfrak{S}_{D'}$ in the sense of preserving inclusion. Then it is proved that $f(\mathcal{R}_D) = \mathcal{R}_{D'}$. Let $D_1,\dots,D_k$ be all 0-simple ideals of S and $D'_1,\dots,D'_{k'}$ be all 0-simple ideals of S'. If $\mathcal{P}(S) \cong \mathcal{P}(S')$, then $k = k'$ and there is a permutation σ on $\{1,\dots,k\}$ such that f induces

$$D_i \cong D_{\sigma(i)} \text{ and } \mathcal{I}_{D_i} \cong \mathcal{I}_{D_{\sigma(i)}} \quad (i = 1,\dots,k).$$

Thus if $\mathcal{P}(S) \cong \mathcal{P}(S')$, then we have $\mathcal{RS}(\mathcal{P}(S)) \cong \mathcal{RS}(\mathcal{P}(S'))$ and $\mathcal{RS}(S) \cong \mathcal{RS}(S')$. By repeated use of nil cover and regular socles and by the similar method as the nil case, we obtain $S \cong S'$. Note that a special treatment is needed for the case where S has a group ideal.

In case $|S| = |S'| = 2$ we can show by the construction of all possible $\mathcal{P}(S)$ that if $S \neq S'$, $\mathcal{P}(S) \neq \mathcal{P}(S')$.

3. POWER SEMIGROUP OF 0-SIMPLE SEMIGROUP

Let $S = \mathcal{M}^0(I,M,G;P)$. Each element X with 0 of $\mathcal{P}(S)$ has the form

$$X = \bigcup_{\lambda=1}^{k} (A_\lambda \times B_\lambda;\ \phi_\lambda)^0$$

where ϕ_λ is a mapping $A_\lambda \times B_\lambda$ into $\mathcal{P}(G)$, i.e. $\phi_\lambda(i,j) \in \mathcal{P}(G)$ for every $(i,j) \in A_\lambda \times B_\lambda$. That is, X is expressed as the union of "rectangles" $(A_\lambda \times B_\lambda : \phi_\lambda)^0$. If $k = 1$, X is called a <u>rectangle</u> element.

For X we define X^* by

$$X^* = \bigcup_{\lambda=1}^{k} (A_\lambda \times B_\lambda)^0.$$

X^* is an element of $\mathcal{P}(S^*)$ where $S^* = \mathcal{M}^0(I,M,\{1\};P^*)$ where P^* is obtained from P by replacing nonzero entries by 1. The union X^* of rectangles $(A_\lambda \times B_\lambda)^0$ is called a <u>normal form</u> if for any pair (λ,μ) of distinct λ and μ, one of the following holds:

(i) $A_\lambda \subset A_\mu$ and $B_\mu \subset B_\lambda$,

(ii) $A_\mu \subset A_\lambda$ and $B_\lambda \subset B_\mu$,

(iii) $A_\lambda \parallel A_\mu$ and $B_\lambda \parallel B_\mu$.

Any finite union of rectangles can be normalized [21]. The minimum of k for which X^* is in a normal form is called the <u>degree</u> of X, denoted by deg X. Define $\|X\|$ by

$$\|X\| = \min \{|\phi_\lambda(i,j)|:(i,j) \in A_\lambda \times B_\lambda,\ \lambda = 1,\dots,k\}.$$

We call $\|X\|$ the <u>depth</u> of X, denoted by dep X.

Let $\mathcal{R}$ denote a $\mathcal{J}$-class in $\mathcal{P}(S)$. We consider $\mathcal{R}$ such that $\mathcal{R}^0$ is 0-simple. The following can be obtained without difficulty.

THEOREM 3.1. [21]. <u>Every element of $\mathcal{R}$ has the same degree and the same depth, and every element of $\mathcal{R}$ contains 0 unless deg $\mathcal{R}$ = 1.</u>

The author has studied idempotents, maximal subgroups and $\mathcal{J}$-classes in $\mathcal{P}(S)$ [21], [22]. Here we report a few results on $\mathcal{J}$-classes which contain idempotents.

Let $X = \bigcup_{\lambda=1}^{k} (A_\lambda \times B_{(\lambda)\epsilon};\ \phi_\lambda)^0$, $Y = \bigcup_{\mu=1}^{k} (C_\mu \times D_{(\mu)\delta};\ \psi_\mu)^0$,

$$X^* = \bigcup_{\lambda=1}^{k} (A_\lambda \times B_{(\lambda)\epsilon})^0, \qquad Y^* = \bigcup_{\mu=1}^{k} (C_\mu \times D_{(\mu)\delta})^0,$$

where ϵ and δ are permutations on $\{1,\dots,k\}$. We arrange these in a normal form, namely

$$A_{\lambda_1} \subset A_{\lambda_2} \text{ if and only if } B_{(\lambda_1)\epsilon} \supset B_{(\lambda_2)\epsilon},$$
$$C_{\mu_1} \subset C_{\mu_2} \text{ if and only if } D_{(\mu_1)\delta} \supset D_{(\mu_2)\delta}.$$

We say X^* is order isomorphic to Y^* or $\{A_1,\ldots A_k\}$ is order isomorphic to $\{C_1,\ldots,C_k\}$ if there is a permutation κ on the set $\{1,\ldots k\}$ such that

$$A_{\lambda_1} \subset A_{\lambda_2} \text{ if and only if } C_{(\lambda_1)\kappa} \subset C_{(\lambda_2)\kappa}.$$

Let $E = \bigcup_{\lambda=1}^{k} (A_\lambda \times B_\lambda;\ \phi_\lambda)^0$, $X = \bigcup_{\lambda=1}^{k} (C_\lambda \times D_\lambda;\ \psi_\lambda)^0$,

and

$$E^* = \bigcup_{\lambda=1}^{k} (A_\lambda \times B_\lambda)^0, \qquad X^* = \bigcup_{\lambda=1}^{k} (C_\lambda \times D_\lambda)^0.$$

Let $\mathbb{R}$ and $\mathbb{L}$ denote the Green's relations. We will state only statements for $\mathbb{R}$.

LEMMA 3.2. Let E^* be an idempotent and X^* an element of $\mathcal{P}(S^*)$. Then $X^* \mathbb{R} E^*$ if and only if X^* is order isomorphic to E^*, $\{C_1,\ldots,C_k\} = \{A_1,\ldots A_k\}$ and there is $Z^* \in \mathcal{P}(S^*)$ such that X^*Z^* is an idempotent which is order isomorphic to E^*.

Let $E_0^* = \bigcup_{\lambda=1}^{k} (A_\lambda^o \times B_\lambda^o)^0$ be an idempotent. Let $\mathcal{Q}_I^*$ be the set of all $\mathcal{A} = \{A_1,\ldots,A_k\}$, $A_\lambda \subset I$ $(\lambda = 1,\ldots,k)$, such that $\bigcup_{\lambda=1}^{k} (A_\lambda \times B_\lambda^o)^0$ is $\mathbb{R}$-related to some idempotent in $\mathcal{P}(S^*)$; and let $\mathcal{Q}_M^*$ be the set of all $\mathcal{B} = \{B_1,\ldots,B_k\}$, $B_\lambda \subset M$ $(\lambda = 1,\ldots,k)$, such that $\bigcup_{\lambda=1}^{k} (A_\lambda^o \times B_\lambda)^0$ is $\mathbb{L}$-related to some idempotent in $\mathcal{P}(S^*)$. Let $\overline{\mathcal{Q}_I}^*$ denote the set of $\overline{\mathcal{A}} = \bigcup_{\lambda=1}^{k} (A_\lambda \times B_\lambda^o)^0$, and $\overline{\mathcal{Q}}_M^*$ the set of $\overline{\mathcal{B}} = \bigcup_{\lambda=1}^{k} (A_\lambda^o \times B_\lambda)^0$. Of course we assume if $\mathcal{A}_1 \neq \mathcal{A}_2$, $\overline{\mathcal{A}}_1$ is not $\mathbb{R}$-related to $\overline{\mathcal{A}}_2$; if $\mathcal{B}_1 \neq \mathcal{B}_2$, $\overline{\mathcal{B}}_1$ is not $\mathbb{L}$-related to $\overline{\mathcal{B}}_2$.

Let $\mathscr{G}(E^*)$ be the maximal subgroup of $\mathscr{P}(S^*)$ containing E^* as the identity element.

$$\mathscr{G}(E^*) = \left\{ \bigcup_{\lambda=1}^{k} (A_\lambda^o \times B_{(\lambda)\pi}^o)^0 : \pi \in \Pi \right\}$$ where Π is a permutation group over $\{1,\ldots,k\}$ such that $\{A_1^o,\ldots,A_k^o\} \to \{B_{1\pi}^o,\ldots,B_{k\pi}^o\}$ is an order anti-isomorphism. Let $\mathscr{R}(E^*)$ denote the $\mathscr{J}$-class containing E^*.

THEOREM 3.3. $\mathscr{R}(E^*) = \overline{\mathscr{Q}}_I^* \cdot \mathscr{G}(E^*) \cdot \overline{\mathscr{Q}}_M^*$.

COROLLARY 3.4. <u>Assume</u> E^* <u>and</u> X^* <u>are idempotents</u>. <u>Then</u> $X^* \mathscr{J} E^*$ <u>if and only if</u> X^* <u>is order isomorphic to</u> E^*.

In the following theorem, we assume E and X are idempotents in $\mathscr{P}(S)$. When $p_{bi} \neq 0$ for some $i \in A_\lambda$, we write $b \,\Delta\, A_\lambda$.

THEOREM 3.4. $X \,\mathbb{R}\, E$ <u>if and only if</u>

(1) $X^* \,\mathbb{R}\, E^*$, <u>(hence</u> $C_\lambda = A_\lambda$ <u>for all</u> λ) <u>and</u>

(2) <u>for each</u> λ <u>and for each</u> $b \in B_\lambda$ <u>with</u> $b \,\Delta\, A_\lambda$ <u>and each</u> $d \in D_\lambda$ <u>with</u> $d \,\Delta\, A_\lambda$ <u>there is</u> $\alpha \in G$ such that

$\psi_\lambda(i,d) = \phi_\lambda(i,b)\,\alpha$ <u>for all</u> $i \in A_\lambda$.

In the similar way as the case of $\mathscr{P}(S^*)$, we construct an $\mathbb{R}$-representative set $\overline{\mathscr{Q}}_I$ and an $\mathbb{L}$-representative set $\overline{\mathscr{Q}}_M$ by using Theorem 3.4 and the result [21] in $\mathscr{J}$-classes of rectangle elements.

THEOREM 3.5. $\mathscr{R}(E) = \overline{\mathscr{Q}}_I \cdot \mathscr{G}(E) \cdot \overline{\mathscr{Q}}_M$.

Let $E = (A \times B;\, \phi)^0$ be an idempotent. If $|\phi(i,j)| = \|E\|$ and $\phi(i,j)$ is a coset of a subgroup H of G, then H is called a <u>ground group</u> of E. It is determined up to conjugacy.

COROLLARY 3.6. <u>Let</u> $E = (A \times B;\phi)^0$ <u>and</u> $F = (C \times D;\psi)^0$ <u>be idempotents</u>. <u>Then</u> $E \mathscr{J} F$ <u>if and only if a ground group of</u> E <u>is conjugate to that of</u> F.

REFERENCES

1. Clifford, A.H. and G.B. Preston, The algebraic theory of semigroups, vol. 1, Math. Surveys No. 7, Amer. Math. Soc., Providence, R.I., 1961.

2. Gould, M. and J.A. Iskra, 'Globally determined classes of semigroups,' Semigroup Forum **28** (1984), 1-11.

3. Kobayashi, Y., 'Semilattices are globally determined,' Semigroup Forum **29** (1984), 217-221.

4. Margolis, S.W. and J.E. Pin, 'Power monoids and finite J-trivial monoids.' Semigroup Forum **29** (1984), 99-108.

5. McCarthy, D.J. and D.L. Hayes, 'Subgroups of the power semigroup of a group,' J. of Comb. Theory, Ser. A, **14** (1973), 173-186.

6. Mogiljanskaja, E.M., 'The solution to a problem of Tamura,' Sbornik Naučnyh Trudov Leningrad. Gos. Ped. Inst. Modern Analysis and Geometry (1972), 148-151 (Russian).

7. Mogiljanskaja, E.M., 'Non-isomorphic semigroups with isomorphic semigroups of subsets,' Semigroup Forum **6** (1973), 330-333.

8. Petrich, M., Introduction to semigroups, Merrill Publ. Co., Columbus, Ohio, 1973.

9. Putcha, M.S., 'On the maximal semilattice decomposition of the power semigroup of a semigroup,' Semigroup Forum 15 (1978), 263-267.

10. Putcha, M.S., 'Subgroups of the power semigroup of a finite semigroup,' Canad. J. Math. **31** (1979), 1077-1083.

11. Spake, R., 'Idempotent free archimedean components of the power semigroup of the group of integers I,' to appear in Mathematica Japonica, 1986.

12. Spake, R., 'The semigroup of nonempty finite subsets of integers,' to appear in Internat. J. of Math. and Math Sci., 1986.

13. Spake, R., 'On the power semigroup of the infinite cyclic group,' Dissertation, University of California, Davis, 1986.

14. Tamura, T., 'The theory of construction of finite semigroups III,' Osaka Math. J. **10** (1958), 191-204.

15. Tamura, T. and J. Shafer, 'Power semigroups,' Mathematica Japonicae **12** (1967), 25-32.

16. Tamura, T. 'On the power semigroup of the group of integers,' Proc. Japan Acad. 60 Ser. A (1984), 388-390.

17. Tamura, T., 'On chains whose power semigroups are lattices,' Semigroup Forum 30 (1984), 35-40.

18. Tamura, T. 'Power semigroups of rectangular groups,' Mathematica Japonica 29 (1984), 671-678.

19. Tamura, T., 'Isomorphism problem of power semigroups of completely 0-simple semigroups,' Journal of Algebra 98, (1986), 319-361.

20. Tamura, T., 'The study of the power semigroup of the group of integers.' (Preprint)

21. Tamura, T., 'The study of the power semigroup of the group of finite 0-simple semigroup I,' (preprint).

22. Tamura, T., 'The study of the power semigroup of finite 0-simple semigroup II,' (preprint).

23. Tamura, T., 'The class of finite semigroups is globally determined.' (In preparation)

TYPE II REDUX

Bret Tilson
P.O. Box 1789
Sausalito, California 94966
USA

INTRODUCTION. Recent work of Ash has shown that the collection of all finite monoids whose idempotents commute is the **M**-variety generated by finite inverse monoids. Extending Ash's ideas, Birget, Margolis and Rhodes have shown that the collection of all finite monoids whose idempotents form a submonoid is the **M**-variety generated by finite orthodox monoids. These results are actually special cases of a more general conjecture, called the type II conjecture.

In 1972, John Rhodes and I introduced the notion of type II elements in a paper called "Improved lower bounds for the complexity of finite semigroups" [LB2]. In that paper, we give a constructive method for determining which regular elements of a monoid are type II. That method lead to the type II conjecture: *The set of type II elements of a monoid* M *is the smallest submonoid of* M *that admits "conjugation".*

The paper [LB2] shows that the type II conjecture holds for regular monoids, and the recent work cited above shows that the conjecture holds for monoids whose idempotents form a submonoid.

Because of the renewed interest in the type II conjecture, I decided to see if the original proof for the regular case (in [LB2]) could be made more accessible. That proof is entangled in the language of Rees matrix representations of regular $\mathcal{J}$ classes, and, in my opinion, obscures the essential ideas and discourages all but the most dedicated from reading it. To my pleasant surprise, this investigation has produced a short, easy, coordinate free proof of the [LB2] result.

I am presenting this new proof in order to clarify the "mysterious" [LB2]. It is my hope that this new proof will spur someone on to solve the general problem.

S. M. Goberstein and P. M. Higgins (eds.), Semigroups and Their Applications, 201–205.

All groups and monoids are assumed finite.

Let M and N be monoids. A relation $\phi: M \to N$ is a ***relation of monoids*** if $\#\phi = \{(m,n): n \in m\phi\}$ is a submonoid of $M \times N$. When, in addition, ϕ is fully defined on M, that is, $m\phi \neq \emptyset$ for all $m \in M$, the relation ϕ is called a ***relational morphism.***

Let M be a finite monoid. An element $m \in M$ is ***type II*** if for every relational morphism

$$(1.1) \qquad \phi: M \longrightarrow G$$

where G is a finite group, we have $(m,1) \in \#\phi$. The set of all type II elements of M is denoted M_{II}. It is easy to see that M_{II} is a submonoid of M.

If S is a semigroup, S_{II} is defined in the same way. However, it is easy to show that

$$S_{II} = S^1_{II} - \{1\}$$

Therefore, I prefer to treat monoids rather than semigroups. Nothing is gained in this case by denying oneself an identity.

The definition of type II elements is existential; it does not lead to a constructive method for finding all type II elements of M. The result in [LB2], presented again here, gives a method of constructing all type II elements of M that are regular in M. Thus if M is regular, then M_{II} can be constructed. The construction method is "conjugation".

A pair (a,b) of elements of M is called a ***conjugate pair*** in M if either $aba = a$ or $bab = b$. If e is an idempotent, then (e,e) is a conjugate pair. Given a conjugate pair (a,b) in M, we consider the operation of ***conjugation*** in M, defined by $m \to amb$.

PROPOSITION 1.1. *Let (a,b) be a conjugate pair in M. Then* $aM_{II}b \subseteq M_{II}$.

Proof. Let $\phi: M \to G$ be a relational morphism, where G is a finite group. It suffices to show that if $m \in 1\phi^{-1}$, then $amb \in 1\phi^{-1}$. Let $(a,g),(b,h) \in \#\phi$. Either $aba = a$ or $bab = b$. If $aba = a$, then

$$(a,g)[(b,h)(a,g)]^k = (a,g(hg)^k)$$

for all $k > 1$. Since G is finite, there exists an n such that $(hg)^n = 1$. Thus $h^{-1} = g(hg)^{n-1}$. It follows that $(a,h^{-1}) \in \#\phi$.

If, on the other hand, $bab = b$, then the same argument shows that $(b,g^{-1}) \in \#\phi$. We may conclude that if (a,b) is a conjugate pair, then there exists $k \in G$ such that $(a,k),(b,k^{-1}) \in \#\phi$.

Now let $m \in 1\phi^{-1}$; in other words, assume that $(m,1) \in \#\phi$. Then

$$(a,k)(m,1)(b,k^{-1}) = (amb,1) \in \#\phi .$$

Consequently, $amb \in 1\phi^{-1}$. □

Therefore, M_{II} admits the conjugation operation over all conjugate pairs in M. Define M_C to be the smallest submonoid of M that admits conjugation. It follows directly that

$$M_C \subseteq M_{II}$$

The "Type II conjecture" is that $M_C = M_{II}$. Since M_C is clearly constructible from the multiplication table of M, the truth of this conjecture would show that membership in M_{II} is decidable. The original type II paper, [LB2], showed that

$$M_C \cap Reg = M_{II} \cap Reg$$

In particular, if M is regular, then $M_C = M_{II}$. An easy proof of this result is now presented.

LEMMA 1.2. M_C *contains the idempotents of* M.

Proof. Let e be an idempotent of M. Then (e,e) is a conjugate pair. Since $1 \in M_C$, we have $e1e = e \in M_C$. □

LEMMA 1.3. *Let* $r,b \in M$ *with* $rb \mathcal{R} r$. *Then there exist* $a \in M$ *satisfying*

$$rba = r \quad \textit{and} \quad aba = a$$

Thus, (a,b) *is a conjugate pair in* M.

Proof. Since $rb \mathcal{R} r$, there exists $w \in M$ such that $rbw = r$. Hence $r(bw)^k = r$ for all $k \geq 1$. Since M is finite, we may choose an $n>1$ so that $(bw)^n$ is idempotent. Set $a = w(bw)^{2n-1}$; then $rba = r$. Now

$$\begin{aligned} aba &= w(bw)^{2n-1}bw(bw)^{2n-1} \\ &= w(bw)^{4n-1} \\ &= w(bw)^{n-1}(bw)^{3n} \\ &= wb(bw)^{n-1}(bw)^{n} \\ &= w(bw)^{2n-1} \\ &= a \end{aligned}$$ □

Let M be a finite monoid and let R be an $\mathcal{R}$-class of M. For $r, r' \in R$, define the relation $r \equiv r'$ if there exists $w, w' \in M_C$ such that $rw = r'$ and $r'w' = r$. Given the $\mathcal{R}$-class R, we may consider the partial transformation monoid (R,M) with action given by right multiplication. We will show that the relation $\equiv$ on R is a right congruence for (R,M). Then we may consider the transformation monoid $(R/\equiv, M)$ with action inherited from right multiplication.

PROPOSITION 1.4 *Let* R *be an* $\mathcal{R}$*-class of a monoid* M. *Then the relation* $\equiv$ *is a right congruence for* (R,M), *and* $(R/\equiv, M)$ *is an injective transformation monoid.*

Proof. Symmetry of $\equiv$ follows directly from the definition. Reflexitivity and transitivity follow from the fact that M_C is a submonoid of M. Therefore, $\equiv$ is an equivalence relation.

To show that $\equiv$ is a right congruence, let $r \equiv r' \in R$ and suppose $b \in M$ with $rb, r'b \in R$. We must show that $rb \equiv r'b$. Since $r \equiv r'$, there exists $w \in M_C$ such that $r' = rw$. Furthermore, since $rb \in R$, Lemma 1.3 implies the existence of $a \in M$ such that (a,b) is a conjugate pair and $rba = r$. From this information we may write

$$r'b = rwb = rbawb$$

Since $w \in M_C$ and (a,b) is a conjugate pair, we have $awb \in M_C$. This shows that there exists an element $z \in M_C$ so that $r'b = rbz$. A dual argument shows the existence of $z' \in M_C$ such that $rb = r'bz'$. Therefore, $rb \equiv r'b$, and $\equiv$ is a right congruence.

Thus, we have shown that $(R/\equiv, M)$ is a tm. To show that this transformation monoid is injective, let $r, r' \in R$ and let $b \in M$ with $rb \equiv r'b$. We must show that $r \equiv r'$. Since $rb \equiv r'b$, there exists $w \in M_C$ such that $r'b = rbw$. Choose $a \in M$ so that (a,b) is a conjugate pair and $r'ba = r'$. Then

$$r' = r'ba = rbwa$$

But $bwa \in M_C$, so we have shown the existence of an element $z \in M_C$ satisfying $r' = rz$. A dual argument establishes $r \equiv r'$. □

LEMMA 1.5. *Let* (Q,M) *be a partial injective tm. If* m *is a type II element of* M, *then* $qm \subseteq q$ *for all* $q \in Q$. *In other words, every member of* M_{II} *must act as a partial identity on* Q.

Proof. Let H be the symmetric group on the set Q. Then

$$\phi : M \longrightarrow H$$

$$m\phi = \{h \in H : qm \subseteq qh \text{ for all } q \in Q\}$$

is easily seen to be a relational morphism. If $m \in M_{II}$, then $(m,1) \in \#\Phi$. It follows that for each $q \in Q$, $qm \subseteq q1 = q$. □

LEMMA 1.6. *Let* R *be a regular* $\mathcal{R}$*-class of* M *and let* $m \in R$. *If* m *acts as a partial identity on* $R/\equiv$, *then* $m \in M_C$.

Proof. Since R is regular, there is an idempotent $e \in R$, and $em = m$. If m acts as a partial identity on $R/\equiv$, then it follows that $e \equiv em = m$. Since $e \equiv m$, there exists a $w \in M_C$ such that $m = ew$. But by Lemma 1.2, every idempotent belongs to M_C, so we conclude that $m \in M_C$. □

Lemmas 1.5 and 1.6 lead us to our destination.

THEOREM 1.7. *Let* M *be a finite monoid and let* Reg *denote the set of regular elements of* M. *Then*

$$M_C \cap \text{Reg} = M_{II} \cap \text{Reg}$$

Proof. Since $M_C \subseteq M_{II}$, it suffices to show that $M_{II} \cap \text{Reg} \subseteq M_C \cap \text{Reg}$. Let m be a regular type II element of M, and let R be the $\mathcal{R}$-class containing m. Since the tm $(R/\equiv, M)$ is injective, Lemma 1.5 states that the element m must act as a partial identity on $R/\equiv$. Lemma 1.6 allows us to conclude that $m \in M_C$. □

This is the result in [LB2].

REFERENCES

[LB2] John Rhodes and Bret Tilson, "Improved lower bounds for the complexity of finite semigroups", *J. Pure Appl. Algebra* 2 (1972), 13-71.

[Ash] C. Ash, "Finite semigroups with commuting idempotents," (preprint, 1986).

[B-M-R] J.-C.Birget, S.Margolis, and J.Rhodes, "Finite semigroups whose idempotens commute or form a subsemigroup," in this volume.

[Rh] John Rhodes, "New techniques in global semigroup theory," in this volume.

INDEX